Maritime Asia

Edited by
Roderich Ptak, Thomas O. Höllmann,
Jorge Flores and Zoltán Biedermann

Volume 28

2015

Centro Científico e Cultural de Macau, I.P.

Harrassowitz Verlag · Wiesbaden

Roderich Ptak and Baozhu Hu

The Earliest Extant
Bird List of Hainan

An Annotated Translation
of the Avian Section in *Qiongtai zhi*

2015

Centro Científico e Cultural de Macau, I.P.

Harrassowitz Verlag · Wiesbaden

Die Reihe *Maritime Asia* setzt die Reihe *South China and Maritime Asia* fort, in der die Bände 1–16 erschienen sind.

The series *Maritime Asia* continues the former series *South China and Maritime Asia*, in which the volumes 1–16 have been published.

Bibliografische Information der Deutschen Nationalbibliothek
Die Deutsche Nationalbibliothek verzeichnet diese Publikation in der Deutschen Nationalbibliografie; detaillierte bibliografische Daten sind im Internet über http://dnb.dnb.de abrufbar.

Bibliographic information published by the Deutsche Nationalbibliothek
The Deutsche Nationalbibliothek lists this publication in the Deutsche Nationalbibliografie; detailed bibliographic data are available on the internet at http://dnb.dnb.de.

For further information about our publishing program consult our website http://www.harrassowitz-verlag.de
© Otto Harrassowitz GmbH & Co. KG, Wiesbaden 2015
This work, including all of its parts, is protected by copyright. Any use beyond the limits of copyright law without the permission of the publisher is forbidden and subject to penalty. This applies particularly to reproductions, translations, microfilms and storage and processing in electronic systems.
Printed on permanent/durable paper.
Printing and binding: Memminger MedienCentrum AG
Printed in Germany
ISSN 1863-6268
ISBN 978-3-447-10315-2

Table of Contents

Preface

Hainan and the regions now called Guangdong and Guangxi have always played a special role in China's past. Government officials, literates, soldiers, merchants and migrants were forced to accept unusual living conditions when moving from the northern plains to the far South. Textual evidence, especially since the Tang period, suggests that Hainan in particular constituted a location which was much feared but also admired for its beauty and exotic setting. In the course of time settlers from the mainland became more familiar with this new world and one encounters a growing number of accounts on various aspects of the island's nature and society. This is a fascinating subject that merits to be studied by sinologists and others, including specialists interested in Hainan's geography and natural history.

Hainan's past has also become a recurrent theme in seminars and research conducted at the Sinological Institute of the Ludwig-Maximilians-Universität in Munich. This involves the investigation of ancient texts and materials related to China's southern provinces and the Nanhai zone more generally, or to maritime China in a larger sense. The present book is one small product emerging from such joint efforts.

As is the case with so many scholarly investigations, it has gone through a complex process of mental metamorphosis before it gradually took its final shape. In fact, the original idea, born several years ago, had very little to do with the final product. At that time the intention was to write a short history of Hainan's natural products and their uses in Song, Yuan and Ming trade. The earliest extant *major* monograph dealing with the island, a *fangzhi* or gazetteer of the sixteenth century, carries long lists of local plants and animals; clearly, such lists are essential for studies of that kind. Other Hainan chronicles contain similar lists with additional data, but until recently local gazetteers were difficult to access; therefore, the original plan was first postponed and then began to disappear...

But somehow, not too long ago, it moved back on stage. By then many more materials related to Hainan's past had surfaced as well, both in printed and electronic form, and a preliminary sighting of these sources led to the conclusion that an investigation of "small issues" should precede a general discussion of larger themes. This in turn fostered the idea that one should look at Hainan's plant and animal lists more carefully.

The very modest study presented here is the direct result of such a "reduced" approach to Hainan's past – or rather, to a much more limited field, namely the history of its avian world. Realistically, it mirrors work in progress; it is like an inflated comment on a short text segment – the bird chapter in a key source related to Hainan under the Ming – and not more than just that. Clearly, one could analyse later chronicles in a similar way or extend the present study to other categories in the same source, for example, to the sections on reptiles and insects, sea animals, herbs, trees, etc.

Work on the present brochure began in early 2014, in association with an advanced research seminar. The intention was to write a joint paper of, say, some thirty pages; but in the course of time the manuscript grew and became more complex than anticipated. We had entered a terminological jungle, compelled to move through a swamp full of traps, before getting back on track. The results are far from being perfect, but there is no methodological ballast, and perhaps the translation we were able to make, along with the many notes and comments, will be of some use for those who wish to follow the same trail.

Finally, we are indebted to Dr. Marc Nürnberger for his technical help. He produced the relevant pdf file of the manuscript before it went off to the publisher. We would like to thank him for his patience and kind support.

R. P. and B. Z. H., October 2014

Introduction

Chinese local chronicles often contain long lists of animals and plants found in a particular region. That also applies to various works related to Hainan and to individual districts located on this island. The lists contained in these accounts are of importance for the natural history of Hainan, which can be treated in its own light, just like the natural history of Taiwan. In both cases the island fauna has its own characteristics and is not completely identical with the one encountered in the coastal regions of the Chinese mainland. The present work is a contribution towards a better of understanding of Hainan's avian world. In essence it is a translation of the earliest extant bird list of Hainan, with notes and long comments, mostly of a philological nature.

This bird list appears in a local chronicle of the Zhengde period (1506–1521). The chronicle bears the title *(Zhengde) Qiongtai zhi* (正德) 瓊臺志 (now *QTZ*) and was written / compiled by a certain Tang Zhou 唐胄 (1471–1539); it is the earliest extant work on Hainan which falls into the "class" of local gazetteers. Earlier Hainan "monographs" available today do not form separate texts; they are usually included in geographical accounts, as special chapters, and often have a strong ethnological touch.

Tang Zhou (style Pinghou 平侯) hailed from Qiongshan 瓊山, a county on northeastern Hainan.[1] He passed the metropolitan examination for the *jinshi* 進士 degree in 1502 and was a respected scholar. But shortly after taking up a government post, his father died and Tang had to return home; this led to his dismissal from office in 1504. He only resumed service for the government in 1522. During the long period in between, from 1504 to 1522, he took care of his aging mother, established a school and taught the young, while collecting local records and literary materials for his own pleasure. He also began to write and gradually finished the *QTZ*.

Back in office, he was repeatedly honoured for his merits: the successful management of flood controls in the Yellow River region, the promotion of local

---

1   For standard biographies of Tang Zhou, see, *Ming shi* 明史, XVIII, j. 203, pp. 5357–5359; Zhu Yihui 朱逸輝, *Hainan mingren zhuanlüe* 海南名人傳略, I, pp. 20–22. There is also a master thesis on Tang Zhou. See, Bu Weibo 補維波, *Tang Zhou yanjiu* 唐胄研究.

agriculture and the establishment of an efficient irrigation system. He also supported the educational sector, especially in remote areas. Once, due to his effort and fame, he made it possible to peacefully resolve a local conflict. However, when suggesting to the court that one should not wage war against Annan 安南 (today: northern Vietnam), his loyalty was no longer appreciated by the Jiajing 嘉靖 emperor; this also had to do with the views Tang had voiced during the so-called "Rites Controversy" (often called *da liyi* 大禮儀).[2] Like so many others, he ended in jail (1538), dying shortly after his release in 1539. In 1567, he received a posthumous title: Right Censor-in-chief of the Censorate (*duchayuan you duyushi* 都察院右都御史).

Here we should return to the *QTZ*, which was prepared after Tang Zhou's first long sojourn on the continent – on the basis of earlier material. Presumably most of these sources circulated on the island of Hainan, in Tang's home county, but possibly he had taken additional books from the mainland back home. Be this as it may, the text of the *QTZ* was finalized towards the end of the Zhengde period; the preface is dated 1521 (Zhengde xinsi 正德辛巳).[3]

The extant version of Tang's *QTZ* is included in the famous Tianyi ge 天一閣 collection.[4] Originally this version had 44 chapters of which four chapters (22–23, 43–44) are now lost.[5] The print of the extant text is very clear, only some

---

2   For this great controversy see, for example, Mote and Twitchett, *The Cambridge History of China*. Vol. 7: *The Ming Dynasty, 1368–1644*, pp. 440–461; *Ming shi jishi benmo* 明史紀事本末, II, j. 50, pp. 733–764; Hu Fan 胡凡, *Jiajing zhuan* 嘉靖傳, pp. 52–108; Zhu Xueqin 朱學勤, *Zhongguo huangdi huanghou baizhuan: Ming Shizong* 中國皇帝皇后百傳：明世宗, pp. 36–82; Goodrich and Fang, *Dictionary of Ming Biography*, I, pp. 315–322.

3   See note 5 here.

4   For a comprehensive study on the Tianyi ge collection: Stackmann, *Die Geschichte der chinesischen Bibliothek Tian Yi Ge*.

5   For the text, see *(Zhengde) Qiongtai zhi* (正德) 瓊臺志, ser. Tianyi ge cang Mingdai fangzhi xuankan, 2 vols. (Shanghai: Shanghai guji shudian, 1964). – Originally, the version found in the Tianyi ge collection did not contain Tang Zhou's preface. But the text of that preface was found in the *(Xianfeng) Qiongshan xianzhi* (咸豐) 瓊山縣志 (preserved in the Shanghai Library) and included, as an annex, in the Shanghai guji reprint of 1964. Here we shall always cite the 1964 version of *QTZ*. – A more recent edition, also in two volumes, but with short characters and modern punctuation, was prepared for the collection *Hainan difangzhi congkan* 海南地方志叢刊, directed by Hong Shouxiang 洪壽祥 and others (published by Hainan chubanshe). – For general appreciations of the *QTZ*, see, for example: Wang Chao 王釗 and Shi Zhenqing 史振卿, "Shi shu Zhengde 'Qiongtai zhi' de xueshu jiazhi" 試述正德《瓊臺志》的學術價值; Situ Shangji 司徒尚紀 and Li Yan 李諺, "Zhengde 'Qiongtai zhi': yi bu jiechu de fangyu zhi zuo" 正德《瓊臺志》：一部傑出的方輿之作. The work is also briefly described in: Franke and Liew-Herres, *Annotated Sources of Ming History*, II, pp. 1021–1022, and in the sources mentioned in the next note, below.

characters remain illegible. The initial chapters contain various maps and tables with valuable details on the geography and administrative development of Hainan.

The avian catalogue, with which we shall deal here, appears in j. 9 (i.e., tu-chan 土產, part *xia* 下) of the *QTZ*. This chapter provides long lists of local products, including many entries on the island's flora and fauna. As was said, comparable lists are also available in many other Ming and Qing gazetteers, but modern studies on such topics are difficult to find and as far as the present authors can tell, there is no comprehensive work on the lists in *QTZ*, at least not in major European languages.

The bird list in *QTZ* contains over fifty names and in many cases offers brief comments on each "species", often combined with short descriptions. Several passages refer to earlier sources some of which are now lost or only known through scattered citations. In the translation below, the different references will be discussed, one by one, and where necessary, compared with other texts of the Ming period, but also with earlier and / or more recent material.

Two major aims are associated with this exercise: (1) one is to identify the ancient bird names in the light of modern taxonomy. In some cases this is difficult or even impossible because there are not enough distinctive features to rely on. But in other cases one can at least define the genus or a small group of similar birds by comparing traditional elements to the ones found in recent zoological catalogues. (2) Several entries in the *QTZ* offer details with a strong cultural dimension. Such observations link the human sphere to the avian world; they mirror religious, ecological, literary and other concerns. Where possible, we shall also comment on that.

The *QTZ* is not the earliest chronicle or *difangzhi* 地方志 of Hainan. Earlier texts, now lost, include several titles briefly presented in modern catalogues such as the one prepared by Li Mo 李默.[6] Some of these works are quoted in the bibliographical sections of traditional texts, as for example in Ruan Yuan's 阮元 *Guangdong tongzhi* 廣東通志 (1822).[7] The following list mentions the items presented by Li Mo:

---

6 See Li Mo, *Guangdong fangzhi yaolu* 廣東方志要錄, pp. 432–435. For extant works one may further consult, for example, Zhang Shitai 張世泰 et al., *Guancang Guangdong difangzhi mulu* 館藏廣東地方志目錄, pp. 139–156, and Wang Huijun 王會均, *Hainan wenxian ziliao jianjie* 海南文獻資料簡介, pp. 137 et seq. A useful study: Wen Huanran 文煥然, "Hainan sheng yixie difangzhi kao" 海南省一些地方志考, pp. 123–125. The collection *Hainan difangzhi congkan* (see previous note) also contains bibliographical information on local chronicles, usually at the end of each volume. For the *QTZ*, see, for example, vol. 2 of that work, pp. 887–890.

7 *(Daoguang) Guangdong tongzhi* (道光) 廣東通志, especially V, j. 192–193, pp. 230–259.

(a) *Zhu Ya zhuan* 珠涯傳 (1 *juan*), compiled by Gai Huang 蓋泓. Evidently this book, possibly a work of the fourth century, was still available in the imperial library of the Sui because it appears in the catalogue of the *Sui shu* 隋書. Scattered references to it may also be found in later texts.[8]

(b) *Qiongzhou tujing* 瓊州圖經, an anonymous text of the Song period, quoted, for example, in *Yudi jisheng* 輿地紀勝 (1221).[9]

(c) *Qiongguan tujing* 瓊管圖經 (16 *juan*), by Zhao Ruxia 趙如厦. This work is listed in the catalogue of the *Song shi* 宋史.[10] Judging by the names, Zhao Ruxia should be related to Zhao Rugua 趙如适, author of the famous *Zhufan zhi* (1225), which also contains a description of Hainan.[11]

(d) *Qiongguan zhi* 瓊管志, a Song work of uncertain authorship, quoted, for example, in *Yudi jisheng*.[12]

(e) *Qiongtai zhi* (title same as *QTZ*), briefly referred to in *Yongle dadian* 永樂大典; the author seems unknown.[13]

(f) *Qionghai fangyu zhi* 瓊海方輿志 (2 *juan*), compiled by Cai Wei 蔡微, listed in Qian Daxin's 錢大昕 *Yuan shi yiwen zhi* 元史藝文志.[14]

(g) *Qiongtai waiji* 瓊臺外紀 (5 *juan*), compiled by Wang Zuo 王佐 (Wang Tongxiang 王桐鄉, 1420–1505), dated 1511; it is listed, for example, in Huang Zuo's 黃佐 *Guangdong tongzhi* (1561).[15]

---

8   *Sui shu* 隋書, IV, j. 33, p. 983; *(Daoguang) Guangdong tongzhi*, V, j. 193, p. 251 top.

9   *Yudi jisheng*, IV, j. 124, 10a–b (pp. 3571–3572); *(Daoguang) Guangdong tongzhi*, V, j. 192, p. 236 bottom.

10 *Song shi* 宋史, XV, j. 204, pp. 5164–5165.

11 The best edition of this work is called *Zhufan zhi zhubu* 諸蕃志注補 (it carries a Chinese translation of the English comments to the translation by Friedrich Hirth und W. W. Rockhill with additional notes by Han Zhenhua 韓振華). Translation: Hirth and Rockhill, *Chau Ju-kua: His Work on the Chinese and Arab Trade*. – The *Qiongguan tujing* is also mentioned, for example, in *(Jiajing) Guangdong tongzhi*, II, j. 42, 19b (p. 1090), as a lost item.

12 *Yudi jisheng*, IV, for example j. 124, 1a–b (pp. 3553–3554), 5b–6a (pp. 3562–3563), j. 125, 1b (p. 3484); j. 126, 1b–2a (pp. 3608–3609), j. 127, 1b et seq. (pp. 3618 et seq.).

13 *Yongle dadian fangzhi jiyi* 永樂大典方志輯佚, IV, pp. 2834–2835. Also see, for example, *(Daoguang) Guangdong tongzhi*, V, j. 192, p. 236 bottom.

14 See *Yuan shi yiwen zhi* 元史藝文志, in *Jiading Qian Daxin quanji* 嘉定錢大昕全集, V, part 2, p. 36. Also *(Jiajing) Guangdong tongzhi*, II, j. 42, 21b–22a (p. 1091), and *(Daoguang) Guangdong tongzhi*, V, j. 193, p. 249 bottom. – For a brief discussion of this lost text: Wen Huanran, "Hainan sheng yixie difangzhi kao", pp. 123–125, and Li Mo, *Guangdong fangzhi yaolu*, pp. 434–435.

Besides the works mentioned by Li Mo, one can still find additional titles cited in other compilations, as for example in the *Yongle dadian*. Some of these titles constitute alternative name forms of the sources mentioned above, but others should be considered as separate works. Not much is known of these texts and they do need to be presented here.[16]

In terms of chronology, the next work in the list would be Tang Zhou's *QTZ* itself. Its bird catalogue contains several citations from texts (f) and (g). Evidently, then, these earlier texts provided data on the fauna of Hainan. Whether the other early titles carried similar information and how the relevant sections, if any, were related to each other, remains largely unclear.

To understand the bird entries in *QTZ* it will be essential to not only consider earlier work, which includes various sources other than *Qionghai fangyu zhi* and *Qiongtai waiji*, but to consult later chronicles as well. In some cases these accounts provide identical or additional descriptions. In other cases, there are fewer details. The Wanli edition of *Qiongzhou fuzhi* 瓊州府志 is a case in point. The bird list in this book contains roughly the same number of entries, but the comments on each bird are shorter.[17] By contrast, Qing works are often very different. They may record alternative names and not infrequently the overall arrangement of the flora and fauna sections varies substantially from the arrangement found in earlier texts. One such case is the Qianlong version of *Qiongzhou fuzhi*, originally prepared by Xiao Yingzhi 蕭應植.[18]

Similar observations apply to a number of Guangdong chronicles, which usually carry long chapters on Hainan. The flora and fauna sections in these gazetteers include data related to both continental Guangdong and the island of Hainan. Among such works one finds the *Guangdong tongzhi chugao* 廣東通志 初稿 (1535), Huang Zuo's *Guangdong tongzhi* (1561), a compilation by Guo Fei's 郭棐 with the same title (1602), and many other texts.[19] The second work in particular has been of help for the present study although one has to use it carefully because it carries a number of inaccuracies. A further text is Qu Dajun's 屈大均 famous *Guangdong xinyu* 廣東新語 (preface 1700), which bears the

---

15 *(Jiajing) Guangdong tongzhi*, II, j. 42, 22a (p. 1091). Also see *(Daoguang) Guangdong tongzhi*, V, j. 193, p. 249 bottom.

16 *Yongle dadian fangzhi jiyi*, IV, pp. 2828 et seq.

17 On Hainan chronicles of the Qing period, see, for example, Wang Huijun 王會均, "Ming xiu 'Qiongzhou fuzhi' yanjiu" 明修《瓊州府志》研究.

18 Li Mo, *Guangdong fangzhi yaolu*, pp. 437, 438. For these and other local chronicles of the Qing periods, which we have used here, see the detailed references in the bibliography.

19 *(Jiajing) Guangdong tongzhi chugao*, j. 31, 17a–18a; *(Jiajing) Guangdong tongzhi*, II, j. 24, 9a–15b (pp. 624–632); *(Wanli) Guangdong tongzhi*, j. 59 (Qiong 3), 27a. For short descriptions of these and later works, again Li Mo, *Guangdong fangzhi yaolu*, pp. 2 et seq.

character of a local almanac or encyclopedia. Generally, this is one of the best-studies Qing sources for Guangdong.[20]

Other accounts considered here include geographical works and accounts that fall into the *lishi dili* 歷史地理 and / or *biji* 筆記 genres. Most of these texts date from the pre-1500 period. Furthermore, the bird list in *QTZ* quotes from the works of famous poets. Some of these verses, so it seems, were already printed in the lost *Qiongtai waiji*. The *QTZ* itself may also have inspired later poets. One example is a long piece by Tang Xianzu 湯顯祖 (1550–1616), which could be based, in part at least, on our text.[21] Finally, there are *bencao* works 本草, ancient dictionaries and books exclusively dealing with animals, besides the "usual" stock of classical sources from Zhou and Han times, some of which are of relevance to the present study.

Here we may briefly switch to modern zoology and the problem of identifying animals. The first "Western" ornithologist to fully describe the avifauna of Hainan in a systematic way, following the conventions and taxonomy then valid in European biology, was Robert Swinhoe.[22] Earlier works in European languages often refer to the avifauna of Hainan, but not in such a systematic way. These texts are not cited here.[23]

After the publication of Swinhoe's study new bird lists have appeared in China and elsewhere, which discuss the avifauna of Hainan and Guangdong, or more generally, of the regions around the Gulf of Tongking. These catalogues have

---

20   The text is included in Ou Chu 歐初 and Wang Guanchen 王慣忱 (eds.), *Qu Dajun quanji* 屈大均全集, 8 vols. (Beijing: Renmin wenxue chubanshe, 1996), but here we use the Zhonghua shuju edition of 1975.

21   Or on Wang Zuo's *Qiongtai waiji*. Wang also left a collection of diverse texts called *Jilei ji* 鷄肋集. This work, which is not very well known, contains verses on birds as well. For these verses one may consult http://www.hnszw.org.cn/data/news/2011/03/48916/ (July 2014). – For Tang Xianzu: *Tang Xianzu shi wen ji* 湯顯祖詩文集, I, j. 11, p. 431: "Haishang zayong ershi shou" 海上雜咏二十首. Also see, for example, Yao Pinwen 姚品文 and Long Xiang-yang 龍祥洋,"Tang Xianzu yu Hainan dao" 湯顯祖與海南島, pp. 77–78.

22   Swinhoe, "On the Ornithology of Hainan". – Swinhoe knew of Chinese local chronicles with data on Hainan's birds. He refers to a text called *Kiung-shan-Heen Che* (p. 90, footnote), certainly one of the extant *Qiongshan xianzhi* 瓊山縣志 versions. For a recent popular analysis of these findings, see http://blog.sina.com.cn/s/blog_65f7afed01011e45.html (June 2014). Also see, for example, Lu Gang 盧剛, "Hainan bainian niaolei diaocha: faxian 16 zhong xin niaolei shang bai zhong niao xiaoshi" 海南百年鳥類調查: 發現 16 種新鳥類上百種鳥消失, on http://www.hinews.cn/news/system/2013/03/18/015533840.shtml; similarly 海南鳥類 300 年記錄史, on http://blog.sina.com.cn/s/blog_63f3c38a0102e7m8.html (both July 2014).

23   One short example in Du Halde, *Description*, I, p. 240.

become more precise over time; the more recent ones are particularly useful when one tries to identify traditional bird names in old Hainanese chronicles.[24] Other works describe the avifauna of China in its totality. Catalogues and studies of this type abound, some were published outside of China. Finally, there are many finely-illustrated surveys and general accounts of individual bird families, or even species. Of all these works, we shall only mention selected items in the footnotes.[25]

The foregoing paragraphs, besides outlining the general background of our text, also reveal certain characteristics that one can associate with the animal lists found in Chinese local chronicles, suggesting how one should deal with them. The following sections provide the translation of the bird segment in *QTZ*; this is followed by a brief conclusion with additional observations and comments.

---

24 Older examples in European languages: Ogilvie-Grant, "On the Birds of Hainan"; Delacour and Jabouille, *Les oiseaux de l'Indochine française*; the same, "Oiseaux des îles Paracels". – Two recent Chinese examples frequently cited in the present study: *Hainan dao de niaoshou* 海南島的鳥獸; Shi Haitao 史海濤, Meng Jiliu 蒙激流 et al., *Hainan luqi beizhui dongwu jiansuo* 海南陸栖背椎動物檢索. – Several modern bird lists are also online. We only list three sources here: http://avibase.bsc-eoc.org/checklist.jsp?region=CNhi&list=howardmoore; http://www.birdlist.org/nas/china/china_south/hainan/hainan.htm; http://www.hnszw.org.cn/data/news/2013/02/56956/ (all May to July 2014).

25 The many works by Zheng Zuoxin 鄭作新 (=Cheng Tso-hsin) have an authoritative character. These include (only some examples): *A Synopsis of the Avifauna of China*; *Zhongguo jingji dongwu zhi: niao lei* 中國經濟動物志: 鳥類; *Zhongguo dongwu tupu: niaolei* 中國動物圖譜: 鳥類; *Zhongguo niaolei xitong jiansuo* 中國鳥類系統檢索; *Zhongguo niaolei zhong he yazhong fenlei minglu daquan* 中國鳥類種和亞種分類名錄大全. – Other fairly recent works (again, examples): Étchécopar et al., *Les oiseaux de Chine, de Mongolie et de Corée non passereaux*; Zhongguo yesheng dongwu baohu xiehui 中國野生動物保護協會, Qian Yanwen 錢燕文, *Zhongguo niaolei tujian* 中國鳥類圖鑒; Zhao Zhengjie 趙正階, *Zhongguo niaolei zhi* 中國鳥類志; Zheng Guangmei 鄭光美 and Zhang Cizu 張詞祖, *Birds in China*. – Among the many general compendia the multi-volume collection *Handbook of the Birds of the World*, edited by Joseph del Hoyo et al., has a special status, but we shall not cite it here. – Recent general Chinese-"Western" dictionaries such as *Le Grand Ricci* (or *Grand dictionnaire Ricci de la langue chinoise*) are very useful as well, but again, they are not listed here.

Translations

Explanatory Note

As was said in the introduction, the translations are based on the Tianyi ge edition of the *QTZ*. However, the segmentation of each entry into phrases follows the modern text in short characters, but there are exemptions and occasionally we use different punctuation marks in the English version.

There are fifty-two entries in all. Some entries offer long explanations with citations from earlier material, including poems (which we turned into prose); others are very brief or provide no detail at all. Here, each entry is arranged in the following way: (1) Chinese text, (2) English translation, (3) English comment.

The comments are limited to essential remarks on cultural aspects, references to key sources and the identification of terms. They mostly address sinologists, but zoologists interested in the history of ornithology may find them useful as well.

For bird species we give the Latin, modern Chinese and English equivalents (where necessary, with variations); for avian families we normally only use the Latin forms. In the notes "primary" sources are cited according to titles; modern works follow a different arrangement: author plus title / short title. The bibliography lists printed works; selected references to internet sources are only given in the notes. Chinese characters, always in the long form, usually appear once, after the first reference to a name / work, etc. They are also in the index.

(1) *Yan* / Swallows and Swifts

燕： 有越胡二種。 越燕卽紫燕， 先春社而來。 東坡《人日》詩云： 新巢語燕還窺硯。 胡燕舊無，弘治間始來巢扵東城門上，社後不歸，今漫衍。《本草》云： 作窠喜長，人言容一匹絹者，令人家富。窠與屎俱堪藥用。又《方輿志》有海燕。

Translation: There are two kinds, the *Yue*[*yan*] 越[燕] and *hu*[*yan*] 胡[燕]. The *Yueyan* is the *ziyan* 紫燕 (literally: purple swallow). It comes before [the time of] the *chunshe* 春社 [festival]. [Su] Dongpo [蘇]東坡 in his poem "Ren ri" 人日 says: "Even the swallow twittering in the new nest peeps [through my window] at the ink-slab." Previously there were no *huyan* [on the island], during the Hong-zhi reign (1488–1505) they first came and nested on the eastern city gate; after the [*chun*]*she* [festival] they did not return; nowadays they have spread [all over]. The *Bencao* 本草 records: They like to build long nests. People say, those which can store one roll of silk, will make a family rich. Both the birds' nest and feces can be used in medicine. [According to the *Qionghai*] *fangyu zhi*, there is [also] the *haiyan* 海燕.

Comment: Robert Swinhoe already refers to a similar entry in a later Hainan chronicle. But in his days it was not yet possible to interpret all the details provided in such sources. Nevertheless, his observations are among the earliest non-Chinese comments on the *yan* birds of Hainan.[1]

Generally, *yan* birds are frequently mentioned in Chinese texts of all kinds. They appear in historical accounts, fiction, drama, poetry and songs. Many relevant descriptions are also accessible through traditional *leishu* 類書 works such as the *Taiping yulan* 太平御覽, compiled by Li Fang 李昉 and others in the early Song period.[2] However, the references collected in *leishu* texts are, for the most part, quite different from the entry in *QTZ* – and also from most other entries encountered in later Hainan chronicles. Indeed, it seems, that these chronicles rely on a different set of sources, i.e., mostly on works related to China's coastal regions, quite in contrast to "conventional" collections that focus on the central and northern parts.[3] Therefore it is necessary to briefly comment on all the descriptive elements found in our text, before drawing a general conclusion.

---

1   Swinhoe, "On the Ornithology of Hainan", pp. 90–91.

2   *Taiping yulan*, IV, j. 922, especially 1a–4b (pp. 4090–4091). – On this important source more generally: Kurz, *Das Kompilationsprojekt Song Taizongs*. – Here, we rarely cite earlier or later *leishu*. Among such works one finds the *Chuxue ji* 初學記 and *Yiwen leiju* 藝文類聚 of the Tang period, as well as the *Gujin tushu jicheng* 古今圖書集成 and many other Qing compilations. All these works, most of which are easily accessible, contain long chapters on birds; by and large the sources they quote are also available through the *Taiping yulan*. The same may be said in regard to certain more specialised titles, for example Lu Dian's 陸佃 (1042–1102) famous *Piya* 埤雅, which cites extensively from the classics and early "Daoist" accounts (for *yan*: II, j. 8, 1a–2a). For general notes on such texts with ornithological sections, see, for example, Taylor, "'Guan, guan' Cries the Osprey".

3   One source quoted in *Taiping yulan* (IV, j. 922, 4b, p. 4091) is the *Guangzhou zhi* 廣州志 (also see farther below, in the present entry here). It mentions three kinds of *yan*. This is later cited, for example, in *(Jiajing) Guangdong tongzhi*, j. 24, 15a (p. 632; wrongly *Guangzhou ji* 廣州記), and *(Daoguang) Guangdong tongzhi*, III, j. 99, p. 249 top.

To begin with, the *QTZ* mentions two kinds of *yan* in its initial phrase: one "from" Yue, purple in colour (*zi* 紫), and one linked to the "Hu". The character *Yue* stands for different ethnic groups in the southern regions, the term *hu* represents the northern "barbarians", especially those living beyond the Great Wall; both expressions may thus be seen to indicate directions.

The *chunshe* festival is celebrated after the start of spring (*lichun* 立春), which is around the day of the spring equinox (*chunfen* 春分); traditionally, a ritual was performed on that day to appreciate the blessings of the earth. Clearly, the observation that *yan* birds would fly in a particular season goes back to very early times. The entry on *yan* birds in *Taiping yulan*, quoted above, gives further references to the links between these birds and the spring season.

Su Dongpo, or Su Shi 蘇軾 (1037–1101), is a famous statesman and poet; his life is well-documented and requires no elaborated comment. Suffice it to say that he was banished to Hainan in 1097, where he wrote several literary pieces complaining about his fate and hoping for his future rehabilitation through the Imperial Court. At the same time he tried to appreciate the island's natural beauty by writing on its exotic flora and fauna. Swallows, he certainly knew, often stay in a particular location for a longer period before returning to their place of origin. Evidently, by making a reference to these birds, Su Dongpo thought to expose his loneliness – as he had done in so many other works.[4]

However, there is one question: the poet was on Hainan in Song times, but our text says swallows first came to the island during the Ming period. There are several possible solutions to this "puzzle": (1) Su Dongpo creates an imaginative setting, because there were no swallows on Hainan in his days. (2) Perhaps a change in weather conditions or other natural phenomena had led to a disappearance of these birds after him; they then came back in the Hongzhi reign. (3) The editors of the *QTZ* committed an error. (4) Su Dongpo and the *QTZ* speak of different birds.

Regarding the "eastern city gate", one cannot tell to which location this refers. There were several towns on the islands. One of the maps in *QTZ* shows

---

4 One English "classic" on Su Dongpo is Lin Yutang, *The Gay Genius*; see especially chapter 27 there. – For the complete poem Gengchen sui ren rizuo... 庚辰岁人日作, see *Su Shi quanji* 蘇軾全集, I, shi ji, j. 43, p. 533. – For a brief description of Su's life in Hainan, see Jordan, "Su Tung-p'o in Hainan"; Hargett, "Clearing the Apertures and Getting in Tune"; Egan, *Word, Image, and Deed*, especially pp. 216–218; Chen Jinxjian 陳金現, "Su Shi zai Danzhou de shenfen rentong" 蘇軾在儋州的身份認同; Pu Youjun 蒲友俊, "Chaoyue kunjing: Su Shi zai Hainan" 超越困境: 蘇軾在海南. – Su wrote many poems about animals. An early study is Xian Yuqing 冼玉清, "Su Shi yu Hainan dongwu" 蘇軾與海南動物, here especially p. 115, where this poem is cited. – Local Hainanese gazetteers also contain much material on Su Shi. The *QTZ* has many references to him, for example in j. 42, 9a–12a.

three gates in the wall surrounding the prefectural city. These gates also appear in the book itself. This includes an eastern gate, but there is no such name on the map.[5]

Bencao works mention the excrements of swallows. But here *Bencao* cannot mean Li Shizhen's 李時珍 (1518–1593) famous *Bencao gangmu* 本草綱目 because this work was compiled much later. Hence, the term or title *Bencao* should point to the *Bencao shiyi* 本草拾遺 or a similar text. Nevertheless, the short description in *QTZ* also occurs in Li's work.[6]

Li Shizhen says that swallows are pungent, bland and poisonous, adding they would be good against worms and devil possession. He also alludes to various stories and folk beliefs connected with them, listing many additional names for swallows, all drawn from earlier material, but this of no importance here. However, he provides a separate chapter on "swallows' nests" as well (entry *huyanke tu* 胡燕窠土) and this of relevance to us.[7]

These "nests", now conventionally called *yanwo* 燕窩, were locally collected in some parts of coastal southern China, especially from offshore islands, and also imported from Southeast Asia. They were a delicassy in the Chinese cuisine and appear under various names, for example *yanchao* 燕巢, *haiyanwo* 海燕窩, etc.[8] There can be no doubt that the *QTZ* refers to such nests in its "quotation" from the *Bencao*, although they have nothing to do with "ordinary" swallows, or *yan*.

Here we should turn to modern zoology. Normally the term *yan* refers to such birds as *Hirundo rustica* (now called *jia yan* 家燕; house swallow, house

---

5   *QTZ*, j. 1, 2b (map), j. 20, 3b (gates).

6   See, for example, the following works / editions: *Chongji Bencao shiyi* 重輯本草拾遺, Qin bu, p. 189 (no. 619); *Bencao jing* 本草經, pp. 302–303; *Bencao gangmu*, IV, j. 48, p. 2634. – Surveys on Li Shizhen, his work and its sources: Unschuld, *Pen-ts'ao. 2000 Jahre traditionelle pharmazeutische Literatur Chinas*, and Nappi, *The Monkey and the Inkpot*.

7   Li Shizhen, *Bencao gangmu*, I, j. 7, pp. 433–434; IV, j. 48, pp. 2633–2634; Read, *Chinese Materia Medica: Avian Drugs*, pp. 52–54 (no. 286). – Among the many names for swallows which one finds in *bencao / leishu* works are *xuanniao* 玄鳥 (because of the bird's dark colour), *yiniao* 乙鳥 (because of its sound), and *zhiniao* 鷙鳥 (because, when flying, it may look like a bird of prey), *xiahou* 夏侯 (for *huyan*), etc. – Li Shizhen's principal source for swallows' nests should be the *Bencao shiyi*; see, for example, *Chongji Bencao shiyi*, Yushi bu, zhong, p. 43 (no. 81).

8   For an overview, see Simoons, *Food in China*, pp. 427–431. – Many details and references, particularly to Yuan works, as well as to Huang Zhong's 黃衷 *Hai yu* 海語, almost a contemporary work of the *QTZ*, in Salmon, "Le gout chinois pour les nids de salanganes". For more recent times, see Chiang, "Market Price, Labor Input, and Relation of Production in Sarawak's Edible Birds' Nest Trade".

swift, barn swallow) or *H. daurica* (also *Crecopis daurica*; *jinyao yan* 金腰燕; golden-rumped swallow, red-rumped swallow) under the *Hirundinidae* family (again *yan* in Chinese). But *yan* is also present in the term *Yuyan ke* 雨燕科, for the *Apodidae*.

We shall look at the *Hirundidae* first. *H. rustica* has been observed in many parts of China and that includes Hainan; it is perhaps the best candidate for a general term like *yan*. The distribution of *H. daurica* is less clear; there are claims that it was recorded on Hainan only recently and that it seems mostly present in the northeastern section of the island and the Wenchang 文昌 region. Could this explain the remark that *huyan* (!) birds started coming to Hainan during the Hongzhi reign? If so, this might also match the assumption that the city gate referred to in the *QTZ* was the one in the prefectural capital, because the capital was then in the northeastern part of the island.

There is a third species under the *Hirundinidae* of Hainan, *Delichon urbica* (*maoqiao yan* 毛脚燕, house martin). This bird was also recorded in more recent times; it is smaller than the other two species.[9]

The purple colour associated with the term *ziyan* in our text remains an open issue. Nevertheless, the implicit equation *ziyan = Yueyan =* modern *jiayan* (plus other popular terms like *sheyan* 社燕) could be correct, depending on the interpretation of the colour pattern.

The second family, the *Apodidae*, is sometimes (but rarely) called *Huyan ke* 胡燕科 and the combination *huyan* already appears in early works. According to the *Erya yi* 爾雅翼 of the Song period *huyan* birds are larger than the *Yueyan*; moreover, they have a strong voice and a white breast with a black pattern.[10] These colours are typical for several members of the *Apodidae* (if one takes white to also include grey) and some of them are indeed slightly larger than the *Hirundinidae*. Modern sources normally list four kinds of *Apodidae* in the context of Hainan: *Hirundapus cochinchinensis* (*huihou zhenwei yuyan* 灰喉針尾雨燕, grey-throated spinetail swift), *Apus pacificus* (*baiyao yuyan* 白腰雨燕, large white-rumped swift), *Apus affinis* (*xiao baiyao yuyan* 小白腰雨燕, house swift), *Cypsiurus parvus* (*zong yuyan* 椶雨燕, palm swift). But one cannot decide to which of these four (or their subspecies) the combination *huyan* in *QTZ* refers (if at all), or whether we should better treat the term *huyan* as a generic name. The last bird, one may add, is mostly found on Hainan and not in other coastal regions of China. This could mean that continental and Hainanese sources are not necessarily dealing with the same kinds when referring to *huyan*.

---

9 See, for example, *Hainan dao de niao shou*, pp. 168–170; Shi Haitao et al., *Hainan luqi bei-zhui dongwu jiansuo*, p. 152.

10 See *Erya yi*, II, j. 15, p. 159.

The final part of the discussion concerns the nests. The ones of the various *Apodidae* and *Hirundinidae* members listed above are unsuitable for the cuisine. Which, then, were the animals providing such nests? Later sources, including the *Guangdong xinyu*, tell us that birds as large as crows built them on small islands near Yazhou 崖州. The *Yazhou zhi* 崖州志, a late Qing gazetteer, cites a passage from the *Guangzhou zhi* 廣州志, which lists three kinds of local *yan*, one being the *tuyan* 土燕. It also quotes from the *Hai yu* 海語 (1534), saying *haiyan* birds as big as *jiu* 鳩 ("pigeons"; or *ya* 鴉, "crows", depending on different editions) would return in spring and nest in caves. People would wait until these birds had disappeared in fall to collect the nests. The *Yazhou zhi* then adds that one can find them on the eastern and western Daimei 玳瑁 islands. These islands, also called Daxiao daimei zhou 大小玳瑁洲, are mentioned in earlier sources as well, for example in *QTZ*.[11] Another, rather amusing reference occurs in a late Guangdong chronicle; it reports that birds would build their nests (*yanwo*) in steep cliffs, beyond the reach of men; therefore tame monkeys (*yuan* 猿) were set free to collect them.[12]

Birds providing edible nests also belong to the *Apodidae* family. But, as was said, this does not involve the species mentioned above. The "nest producers" include, for example, *Aerodramus fuciphagus* (also *Collocalia fuciphaga*; *Zhaowa jisi yan* 爪哇金絲燕 in modern Chinese works; usually "edible-nest swiftlet" in modern English) and *A. maximus* (*C. maxima*, *da jinsi yan* 大金絲燕, black-nest swiftlet). They are mostly distributed in Southeast Asia, where swiftlet farming has led to many discussions and research.

A. fuciphagus has been reported on Dazhou Island 大洲島 off Hainan. Some authors define this kind as a subspecies, calling it *A. f. germani*, a name usually associated with Oustalet (1878), others refer to *A. germani*.[13] Perhaps the *QTZ*, dating from much earlier times, also means this bird when briefly mentioning the term [*yan*]*wo*. The status of *A. maximus* is less clear; one finds the black-nest swiftlet along the China coast, but there are no Ming reports on collecting its nests from Hainan.

Finally there is the term *haiyan* in our text. Today this name occurs in connection with birds under the *Hydrobatidae* (storm petrels); these have nothing to

---

11   *Guangzhou zhi* (third century) as quoted in Li Fang, *Taiping yulan*, IV, j. 922, 4b (p. 4091); *Hai yu*, j. 2, 4b; *Guangdong xinyu*, j. 14, p. 391; *Yazhou zhi* 崖州志, j. 4, p. 79. – Also Salmon, "Le gout chinois pour les nids de salanganes", p. 256. – Note: Some Qing chronicles pertaining to other regions of Hainan repeat the information given in *Hai yu*, but these works are not listed here. – For Daxiao daimei zhou, see *QTZ*, j. 6, 11a.

12   *(Daoguang) Guangdong tongzhi*, III, j. 99, p. 249 top.

13   Zheng Zuoxin, *A Synopsis of the Avifauna of China*, p. 348, Interestingly modern bird lists dealing with Hainan do not always list *A. fuciphagus*.

do with the birds mentioned above. But since the *QTZ* gives no further details in regard to the *haiyan*, one cannot really specify the bird possibly meant by this combination.[14]

To summarize: There are good reasons to assume that *Yueyan* should refer to *H. rustica*, while one cannot tell whether *H. daurica* and / or *D. urbica* were already present on Hainan in these early days and recognised as different birds. *Huyan* is likely to stand for one kind of swift or for several *Apodidae* birds. The reference to [*yan*]*wo* is added to the entry in *QTZ* simply because of the name element *yan*. Finally, swifts and swallows are similar in appearance; this is certainly one reason for their repeated confusion in ancient texts.

(2) *Boge* / Pigeons or Doves

鵓鴿：一曰舶鴿，色興鴉皀，銀灰次之。懸哨夜放者尤尚。

Translation: Another name [for this bird] is *boge* 舶鴿 (literally: boat-pigeon). Its color is largely black, like that of a *ya* 鴉 (crow); the silver grey ones come second. Releasing [a pigeon] with a whistle [tied to its body] during the night [can be] of great advantage.

Comment: Both the single character *ge* 鴿 and the combination *boge* usually stand for pigeons. Pigeons can be raised domestically. They are considered docile and sailors in different parts of the world often took tame pigeons on board their ship in case they would need to deliver messages to their families at home.[15] This also seems to be evoked by the term "boat-pigeon" in our text, unless we should read 舶 as a phonetic variation of 鵓. The character *xing* 興 in the second phrase of the *QTZ* entry is rather strange; it could be a verb or a mistake for *yu* 與; the translation "largely" is free.

The last sentence of the *QTZ* alludes to an event recorded in the context of hostilities between Song and Xi Xia (1038–1227) in 1141: Xi Xia prepared several pigeons, hidden in boxes, which fell into Song hands. When the boxes

---

14 One may add, however, that some later texts dealing with Hainan compared their size to *jiu* 鳩 (usually doves); see *(Minguo) Gan'en xianzhi* (民國) 感恩縣志, j. 4, pp. 87–88; *Yazhou zhi*, j. 4, p. 79; *(Minguo) Danxian zhi* (民國) 儋縣志, j. 3, p. 190.

15 In the Tang period Duan Chengshi 段成式 (?–863) refers to pigeons on board of ships from Bosi 波斯 ("Persia" / Islamic merchants with West Asian origins). See his *Youyang zazu* 酉陽雜俎 (c. 875), j. 16, p. 87 bottom. – More on pigeons in *Kaiyuan Tianbao yishi shi zhong* 開元天寶遺事十種, p. 69, or *Bamin tongzhi* 八閩通志 (originally 1491), I, j. 25, p. 724. – For a short summary of pigeon-raising in China, see Guo Fu 郭郛 et al., *Zhongguo gudai dongwuxue shi* 中國古代動物學史, pp. 428–430.

were opened, the pigeons flew up and the Xi Xia observers could guess how far the Song troops had advanced. This contributed to a decisive Xi Xia victory over Song.[16]

Pigeons were not only used as signal birds in war, raising them was a civilian pastime as well. Qu Dajun has a long chapter on the *ge* 鴿, which presents details in regard to Guangdong, but he does not mention Hainan. There is also a Qing work called *Ge jing* 鴿經.[17] The Daoguang version of *Guangdong tongzhi* says large pigeons were called *baidi* 白地, literally "white earth", because they could not fly; at the same time the term *baige* 白鴿 was in use for all *ge* in Guangzhou.[18] In another source *baige* is equated with 鵓鴿.[19] Again, this could point to a phonetical relation between these two and 舶鴿 in both Mandarin and Cantonese.

Pigeons also appear in poetry; one famous piece is the "Xun ge fu 馴鴿賦" by Wang Shizhen 王世貞 (1526–1590).[20] Later Hainan chronicles contain additional data that are sometimes listed under *ge*. But details reported in these texts often deviate from the ones recorded in *QTZ*; moreover, the qualities are graded differently as well.[21]

Today, there are several kinds of pigeons on Hainan all of which belong to the *Columbidae* family, now called *Jiuge ke* 鳩鴿科.[22] The character *jiu* in that term forms a separate entry in *QTZ* and we shall return to it farther below; it may stand for pigeons but for other birds as well. Normally *jiu* carries an "attribute",

---

16 See *Song shi*, XL, j. 485, p. 13997, where the term *xuanshao jiage* 懸哨家鴿 appears. There are also references to such "signal pigeons" in several other contexts. See, for example, http://www.bjguoxue.com/bjgx/mingrenshuo/2255.jhtml *and* http://m.wangchao.net.cn/bbs/tcdetail_739420.html (both July 2014).

17 *Guangdong xinyu*, j. 20, pp. 527–528. – For Zhang Wenzhong's 張萬鍾 *Ge jing*, see Siebert, *Pulu*, p. 244 n. 446, p. 282. For related texts with explanations: *Mingdai Ge jing Qingdai Ge pu* 明代鴿經 清宮鴿譜. A frequently cited history of pigeon-raising in China: Xie Chengxia 謝成俠, "Zhongguo yang ge de lishi" 中國養鴿的歷史.

18 See *(Daoguang) Guangdong tongzhi*, III, j. 99, p. 249 bottom, quoting earlier material.

19 See *(Xuantong) Ding'an xianzhi* (宣統) 定安縣志, j. 1, p. 104, and *(Guangxu) Ding'an xianzhi* (光緒) 定安縣志, j. 1, p. 130.

20 For a short biography of Wang Shizhen, see Goodrich and Fang, *Dictionary of Ming Biography*, II, pp. 1399–1404. The piece is in his work *Yanzhou xugao* 弇州續稿, j. 1, 12a–13a (p. 7).

21 See, for example, *Yazhou zhi*, j. 4, p. 83: *ge* have red feet; some birds are grey and white, others have black patches; one can distinguish them by eye colour; etc. Similar in *(Mingguo) Gan'en xianzhi*, j. 4, p. 91.

22 See, for example, *Hainan dao de niao shou*, pp. 113–123; Shi Haitao et al., *Hainan luqi beizhui dongwu jiansuo*, pp. 131–133.

which has led to much terminological confusion, as we shall see in the comments to entries 12, 19 and 52. Other traditional expressions for pigeons / doves include combinations such as *huzhou* 鶻鵃.[23]

In all these traditional cases it is practically impossible to define a particular species. Nevertheless, the reference to a grey-coloured kind could point to *Streptopelia orientalis* (*shan banjiu* 山斑鳩, Oriental or rufous turtle dove), *Oenopopelia tranquebarica* (also *S. tranquebarica*; *huo banjiu* 火斑鳩, red turtle dove), or even *Ducula badia* ([*shan*]*huang jiu* [山]皇鳩, [mountain] imperial pigeon). The black coloured kind is less easy to identify, because several species have black parts, while none looks entirely black like a crow.

(3) *Maque* / Sparrows

麻雀：糞卽白丁香，治眼翳。

Translation: [Its] feces is [what one calls] *bai dingxiang* 白丁香; [this can be used for] treating eye diseases.

Comment: The term *dingxiang*, without the attribute *bai*, usually refers to cloves. In ancient times cloves came from the Moluccan Islands and were carried to China, India, the Near East and even the Mediterranean world. The expression *bai dingxiang* is different; it often refers to *excrementum passeris* and one finds it in *bencao* works such as the compendium by Li Shizhen, which discusses its uses in medicine; that includes applications against certain eye problems.[24]

Maque birds are also called *waque* 瓦雀 (literally: tile-sparrow).[25] The latter appears farther below in our bird list (see no. 26); we shall discuss this term separately. The translation "sparrow" is vague. The reason is very simple: *maque* can refer to a large variety of species. The same applies to the single character *que*, which usually designates small birds appearing in flocks.[26] Nevertheless,

---

23 An old study on pigeons and doves in traditional Chinese contexts is Watters, "Chinese Notions about Pigeons and Doves"; see especially pp. 226–229 for the *ge*.

24 For cloves: Ptak, "China and the Trade in Cloves, circa 960–1435", and "Asian Trade in Cloves circa 1500: Quantities and Trade Routes". – Also see *Bencao gangmu*, especially III, j. 34, pp. 1940–1944 (*dingxiang*); IV, j. 48, p. 2629–2930 (under *xiongqueshi* 雄雀屎); Read, *Chinese Materia Medica: Avian Drugs*, p. 51 (no. 283).

25 See, for example, *(Qianlong) Qiongzhou fuzhi* (乾隆) 瓊州府志, j. 1 xia, 98a (p. 103); *(Minguo) Gan'en xianzhi*, j. 4, p. 89; *Yazhou zhi*, j. 4, p. 81; *(Minguo) Danxian zhi*, j. 3, p. 191; *(Guangxu) Ding'an xianzhi*, j. 1, p. 131.

26 A common explanation is this: *que* combines two elements, *xiao* 小 (rad. 42) for a small bird, and *zhui* 隹 (rad. 172) for birds with short tails. See, for example, Read, *Chinese Materia Medica: Avian Drugs*, p. 49 (no. 283).

some scholars have equated *maque* with *Passer montanus* (*shu maque* 樹麻雀, tree sparrow).

Many bird catalogues subordinate *P. montanus* to the *Ploceidae* (*Wenniao ke* 文鳥科). Other works deviate from this convention: the category *Ploceidae* is defined as *Zhique ke* 纖雀科, and *P. montanus* enters the *Passeridae* (*Que ke* 雀科; also used for *Fringillidae*).

P. montanus, *P. rutilans* (*shan maque* 山麻雀; cinnamon sparrow, Eurasian tree sparrow) and other "sparrow-like" birds are very common in southern China, *P. domesticus* (*jia maque* 家麻雀, house sparrow) is at home in the northern regions. The first one, mostly brown in colour, also lives on Hainan.[27] Perhaps, then, the *QTZ* refers to *P. montanus*, but the same could be said in regard to the *waque*. Alternatively, both may refer to different birds, or *maque* should be read as a generic term, while *waque* is more specific. Here, we opted for a general solution, suggesting that both expressions were (and are) largely interchangeable.

One observation should be added: The *Guangdong tongzhi chugao* has the form *maque* 麻鵲 in lieu of 麻雀. But the sequence of the bird list in that text is the same as in *QTZ*, at least for items 2 to 7: *boge, maque, xique, shanhu, ying-wu, quyu*.[28] This suggests that the authors / compilers of that book probably took the *maque* for a close "relative" of the *xique* 喜鵲 (the next item), or a kind of *que* 鵲 more generally.

(4) *Xique* / Magpies

喜鵲： 舊無。景泰初，指揮李翊自高化取雌雄十餘，縱之城隍間。迄今孳育，散至近縣間有之。唐冑詩：

> 橫成銀漢仙橋遠，派衍金籠海國賒。
> 地氣北來知世運，喳喳傳喜遍天涯。

Translation: Formerly there were no [*xique* on Hainan]. In the early Jingtai reign (1450–1456) Commander Li Yi 李翊 took more than ten pairs from Gao-Hua 高化 [to Qiongzhou], which he released near the [local] Chenghuang 城隍 [temple]. Until today they have multiplied and are now scattered throughout the district. Tang Zhou's poem [reads]:

> Stretching across the Milky Way, the fairy bridge [seems] distant,

27   See, for example, *Hainan dao de niao shou*, pp. 272–273; Shi Haitao et al., *Hainan luqi bei-zhui dongwu jiansuo*, pp. 172–173.

28   (*Jiajing*) *Guangdong tongzhi chugao*, j. 31, 17b.

Leading off the golden cage, the maritime kingdom is far away.
Earthly breath moves in from the North, they know the fate,
[Their] twitterings spread good news, as far as Heaven's edge.

Comment: The *QTZ* mentions several Chenghuang temples. As elsewhere in China, these temples were dedicated to the local city god. One such temple is shown on the *QTZ* map of Hainan's prefectural capital.[29]

Different entries in the *QTZ* confirm that Li Yi was a *zhihui* (*qianshi*) 指揮 (僉事), usually translated as "(assistant) commander", who became involved in military actions against bandits and also opened a well for the island's inhabitants. He hailed from Hefei 合肥 (in modern Anhui) and moved to Hainan in 1436.[30] Later chronicles are less precise, saying he came from *haibei* 海北, i.e. from the area north of the Qiongzhou Strait 瓊州海峽, which separates Hainan from the Leizhou peninsula 雷州半島. Other sources associate Li Yi with Huazhou 化州.[31] This last location is present in the combination "Gao-Hua", which refers to Gaozhou 高州 *and* Huazhou, both in western Guangdong (roughly the region of modern Maoming 茂名). Elsewhere in the *QTZ* one finds the sequence "Gao Hua zhu lu" 高化諸路.[32] The administrative term *lu* was current under the Yuan. In 1368, when the Ming took over, they turned Gaozhou *lu* and Huazhou *lu* into *fu* 府, or prefectures.

Whether Li Yi really introduced *xique* birds to Hainan (also see the next entry, no. 5) remains an open issue. Tang Xianzu 湯顯祖, a man familiar with the South and one of China's famous poets and playwrights, knew about that story and later gazetteers repeat it as well.[33] But there are doubts because Su Shi, writing much earlier, in the Song period, had already mentioned the *xique* in his poetry.[34]

---

29 *QTZ*, j. 1, 2b.

30 There are several references to Li Yi in *QTZ*; see, for example j. 5, 10a; j. 18, 6b–7a; j. 19, 7b; j. 21, 8a. – Most likely Li Yi's position implied rank 4a.

31 See, for example, *(Kangxi) Changhua xianzhi* (康熙) 昌化縣志, j. 3, p. 48. – Swinhoe, "On the Ornithology of Hainan", p. 351, also refers to Li Yi and his magpies.

32 For example, *QTZ*, j. 18, 7a.

33 See *Tang Xianzu shi wen ji*, I, j. 11, p. 431. The relevant section appears in "Haishang zayong ershi shou" already cited in the introduction, above. This text contains references to many other birds, besides short phrases similar to the ones found in *QTZ*. There are various works on Tang Xianzu and southern China, including Hainan. See, for example, Fan Zhouyou 範舟游 and Gong Zhongmo 龔重謨, "Tang Xianzu zai Lingnan" 湯顯祖在嶺南", http://txz.suichang.gov.cn/txzyj/qtlw/200708/t20070831_88299.htm (June 2014).

34 This follows Xian Yuqing, "Su Shi yu Hainan dongwu", p. 114. The poem is in *Su Shi quanji*, I, j. 41, p. 514–515. It should be related to Hainan because it occurs in a cycle of poems all

Modern zoological works dealing with Hainan associate the traditional name *xique* with the scientific term *Pica pica*, i.e., the common or Eurasian magpie under the *Corvidae* family.[35] This bird is also widely distributed across continental China and on Taiwan. It is often confused with *Cyanopica cyana* (*hui xique* 灰喜鵲, the azure-winged magpie), which only occurs in the central and northern parts of China, but not on Hainan. Some traditional works refer to the latter under the name *shan xique* 山喜鵲. Both *P. pica* and *C. cyana* also appear on paintings and other objects of art.[36]

Unfortunately the *QTZ* provides no description of the *xique*'s appearance; theoretically, then, the text above could also refer to a different bird, with a colour pattern similar to that of *P. pica* and perhaps with a similar shape. Suitable candidates are found under the genus *Cissa* (or *Urocissa*); this includes, for example, *C. whiteheadi* (*hui lanque* 灰藍鵲 or *baichi lanque* 白翅藍鵲, White-head's blue magpie or white-winged magpie) and *C. erythrorhyncha* (*hongzui lanque* 紅嘴藍鵲, red-billed blue magpie). *Crypsirina formosae* or *Dendrocitta formosae* (*hui shuque* 灰樹鵲, grey tree pie) could be a further option. By contrast one should perhaps exclude *Crypsirina temnura* or *C. temia* (*panwei shuque* 盤尾樹鵲, ratched-tailed or racket-tailed tree pie), because this bird has a very unusual tail; moreover it is restricted to Hainan, therefore it cannot classifiy as a "Ming import"[37]

A further possibility is this: Su Shi thought of one bird, Li Yi took a different kind to Hainan. *Cissa whiteheadi*, *C. erythrorhyncha*, *Crypsirina formosae* and *P. pica* are all found in modern Guangdong and Guangxi. *C. erythrorhyncha* in particular looks similar to *P. pica*; perhaps both were confused, or one was imported, while the other belonged to the local fauna since very early times. The conclusion is that *xique* is likely to stand for *P. pica*, but we chose the translation "magpie" to also indicate a possible semantic extension of that term.

Here we may briefly turn to Tang Zhou's poem. The *QTZ* contains many verses by this man. The opening lines of the present verses evoke the idea of long distance and space. They can be linked to the story of the two loving spirits or stars known as Zhinü 織女 (in the Lyra constellation) and Niulang 牛郎 (in Aquila), separated by the Milky Way (also Yinhe 銀河). The magpie, so the story goes, makes it possible to "bridge" the enormous distance between these

---

of which are linked to that island. Nevertheless, the name Jiangling 江陵 in the title suggests an imagined world outside of Hainan.

35   See, for example, *Hainan dao de niao shou*, pp. 207–208, which also cites the *QTZ* passage.

36   See, for example, *Gugong niao pu* 故宮鳥譜, I, pp. 34 et seq. (there, also similar birds like *Cissa erythrorhyncha*, then called *shanzhe* 山鷓). The *Gugong niao pu* dates from 1761.

37   See, for example, *Hainan dao de niao shou*, pp. 206–210.

stars. Related verses occur in many literary works, for example in a frequently cited poem by Yang Yi 楊億 (974–1020).[38]

Lines three and four suggest positive influences from the North, i.e. from the mainland. The third line associates magpies with changing *qi* 氣 and the birds' ability to foresee the future. *Xi* 喜, in the fourth line, clearly alludes to the *xique* bird, which is a common symbol of good fortune. The last two characters – *tianya* 天涯 – evoke a famous location at the southern extremity of Hainan, in modern Sanya 三亞; today this place is widely known as Tianya haijiao 天涯海角.[39] The implicit political dimension could be this: Magpies spread happiness from the mainland to the remotest corner of Hainan. Put differently, Li Yi contributed to local well-being and closer relations between the island and the continent.

(5) *Shanhu* / Black-throated Laughing Thrushes

山呼：黑腮紫毛者多，純白間有。土人養鬭者多尚白臉，出海北，腮白毛藍。 呼本地爲鐵臉。《外紀》：稍隔一望之海， 而物産斷不相入者有三：虎也，喜鵲也，山呼也。喜鵲自指揮李翊取放後，至今五十餘年，雖間有之，然常落落如遠鄉覊旅，有失路恓惶狀。曾有好事者，取海北山呼放諸園林，即爲火鷄取食，竟莫能育。氣候不同，物情不相入，有如此。

Translation: The ones with black cheeks and purple feathers are common, occasionally there are purely white ones [as well]. The ones which the natives raise for bird fights mostly have a white face and come from the *haibei* [side]; [their] cheeks are white [and their] feathers blue. The native [*shan*]*hu* has a face [with] iron [colour]. The [*Qiongtai*] *waiji* [reads]: "[Both sides of the Qiongzhou Strait] are at eyesight distance, but there are three 'local products' in which they differ: tigers, magpies and *shanhu* [birds]. It has been more than fifty years since Commander Li Yi set free [his] magpies; although one occasionally [sees] them, they are scattered here and there, like distant travellers who lost their way and became perplexed. Once busybodies took *shanhu* [birds] from the *haibei* [side to Qiong-

---

38 Yang Yi, "He hua" 荷花, line Yinhe qiao heng que 銀漢橋橫鵲. See *Quan Song shi* 全宋詩, III, j. 120, p. 1404.

39 However, the combination "Tianya haijiao", strictly speaking two locations, was not always associated with southern Hainan. See, for example, *Lingwai daida jiaozhu* 嶺外代答校注 (text originally 1178), j. 1, pp. 38–40; Netolitzky, *Das Ling-wai tai-ta von Chou Ch'ü-fei*, pp. 17–18 (names of two pavilions in Qinzhou 欽州, on the mainland). Tianya itself occurs in other contexts as well; for instance, as the name of a mountain in Shanxi.

zhou], which they set free in [various] parks and woods; as they were immediately eaten by *huoji* 火鶏 birds, one could no longer raise [them]. The climate is different [on both sides of the strait and] the nature of things also differs; it is like that."

Comment: Li Yi appears in the previous entry (no. 4). There were no tigers on Hainan; elsewhere the *QTZ* confirms this observation.[40] The term *huoji* poses many problems; the *QTZ* carries a special entry on this bird and we shall return to it below (see entry no. 30). It is not clear where the quotation from *Qiongtai waiji* ends; later Hainan chronicles do not cite these lines.

The most impressive point in the text is that the author (of *QTZ* or the *waiji*) notices the climatic effects on the local fauna of a particular region, even though, as in the present case, the distance between the two areas in question – across the Qiongzhou Strait – is very short. Simply put, the environmental setting largely determined the characteristics each species would develop in the course of time to ensure its survival.

The expression *shanhu* can also be written differently: *shanhu* 珊瑚, *shanwu* 山烏, and *shanhu* 山鶘.[41] Some traditional sources list the *shanhu* together with another bird – the *huamei* 畫眉. The *QTZ* carries a separate entry on the latter (see no. 24). Generally, both birds were praised for their fighting qualities. One long entry, entitled *shanhu huamei* 山鶘畫眉, appears in *Guangdong xinyu*.[42] It tells us how these lively animals were kept and fed, and how one should train them to fight.

According to this source, *shanhu* birds were green and purple. The ones with iron (black) feet and red eyes were excellent fighters. Those with some black feathers on their chest and certain other features were good at singing. An earlier source, Huang Zuo's *Guangdong tongzhi*, says they had purple feathers; occasionally there were white ones as well, with a clear voice. There then follows a long story which also mentions black-feathered birds. The colour attributes in *Guangdong tongzhi* are the same as in *QTZ*, but Huang's text, just as the *Guangdong xinyu*, refers to continental Guangdong and not necessarily to Hainan. Regarding continental Guangdong, one can find many more entries in other works, often with different details. The *Sancai zaoyi* 三才藻異 by Tu Cuizhong 屠粹忠

---

40   *QTZ*, j. 9, 1a, quoting *Han shu* 漢書, VI, j. 28 xia, p. 1670.

41   See, for example, sources in the next two notes as well as, for example: *(Minguo) Gan'en xianzhi*, j. 4, p. 90; *(Qianlong) Qiongzhou fuzhi*, j. 1 xia, 97a (p. 103); *Yazhou zhi*, j. 4, p. 82; *(Guangxu) Ding'an xianzhi*, j. 1, pp. 130–131; *(Minguo) Danxian zhi*, j. 3, p. 189.

42   *Guangdong xinyu*, j. 20, pp. 517–518. Similar information in *Nanyue biji*, j. 8, 6b–7a, 8b (but separate entries).

(1629–1706), a work that is rarely cited, says, for example, the ones in Lianzhou 廉州 (Guangxi) show patterns in white and black.[43]

Regarding the *shanhu* of Hainan, one finds an entry on these birds in the Wanli edition of *Qiongzhou fuzhi*. This entry is very brief and repeats some of the information given in *QTZ*, but without quoting the *Qiongtai waiji*. The 1774 version of *Qiongzhou fuzhi* is different: besides confirming the belligerent nature of the *shanhu*, it says they were red and green, with some white parts, and also kept by many people.[44]

The above raises several questions: The *QTZ* refers to imported "blue-and-white" fighting birds kept in cages. When exposed to the wild, these animals would not survive. However, imports of cage birds may have continued in later periods, although we were unable to find references to such imports in Qing works. Moreover, it is not exactly clear, which birds are meant.

Apparently the second kind of *shanhu* mentioned in *QTZ* was a native of Hainan. The colours of these animals – black / purple, sometimes white – vaguely match the ones given in the Guangdong chronicles. By and large these descriptions should mostly refer to one or several birds under the genus *Garrulax*. *Garrulax chinensis*, i.e., the black-throated laughing thrush, a species widely distributed across southern Guangdong, Guangxi and Hainan, is a candidate. Its many modern Chinese names, for example *heihou saomei* 黑喉噪鶥, *heihou xiaodong* 黑喉笑鶇 or *heimian xiaodong* 黑面笑鶇, suggest a dark plumage and extraordinary vocal talents.[45]

Modern zoological accounts specify a Hainanese variety of that bird, namely *G. c. monachus*; they point out that there is much colour variation within the species, just as in the case of *huamei* birds (see below). These variations could be responsible for divergent descriptions in old texts. Moreover, perhaps they should allow us to accommodate the blue-and-white variety under the general "cluster" of *G. chinensis*, but of course that could be wrong as well. Finally, other *Garrulax* species also show black, grey, reddish and / or white parts. This includes, for example, *G. pectoralis* (*heiling saomei* 黑領噪鶥, greater necklaced laughing thrush), *G. canorus* (*huamei* 畫眉, hwamei), *G. monileger* (*xiao heiling*

---

43 *(Jiajing) Guangdong tongzhi*, II, j. 24, 11b (p. 630); *Sancai zaoyi*, j. 8, 69b (p. 57). For illustrations of different kinds of *shanhu* (*shanwu*), see *Gugong niao pu*, II, pp. 68–73. Attention: one of these birds was identified as *Cissa chinensis*.

44 *(Wanli) Qiongzhou fuzhi* (萬曆) 瓊州府志, j. 3, 98a–b (p. 76); *(Qianlong) Qiongzhou fuzhi*, j. 1 xia, 97a (p. 103). The entry in the Wanli version of that text is much shorter.

45 See, for example, *Hainan dao de niao shou*, pp. 233–234; Shi Haitao et al., *Hainan luqi beizhui dongwu jiansuo*, p. 164. In recent years the taxonomy around the *Garrulax* cluster was exposed to many discussions and changes. Some works subordinate these birds to the *Timaliinae* subfamily, others place them under the *Timaliidae* family, or under the *Leiotrichidae*.

saomei 小黑領噪鶥 or *ling xiaodong* 領笑鶇; lesser necklaced laughing thrush), and others.[46] Therefore, one cannot totally exclude the possibility that these birds were occasionally confused with *G. chinensis*. The reference to a purely white kind is less easy to explain; this could refer to a completely different bird.

(6) *Yingwu* / Red-breasted Parakeets

鸚鵡：産西路。俗傳雛中有觜黑不變者名墨賴，不能言。丘濬詩：

爲禽只合作禽言，水飲林棲任自便。
只爲性靈多巧慧，一生常是被拘牽。

Translation: They come from the [area along the] "Western Route" (xilu 西路). Following tradition the ones among the small birds, which have a black bill that does not change, are called *molai* 墨賴; they cannot speak. A poem by Qiu Jun 丘濬 [says]:

For a bird it is suitable to speak the bird language,
To drink water and rest on a tree for convenience.
Just because it is clever by nature, skilful and intelligent,
Its life is often [subjected] to captivity and manipulation.

Comment: "Western Route" should refer to the western side or coast of Hainan.[47] Qiu Jun (1421–1495) was exiled to Hainan. He is famous for his literary works and has left several poems, some of which are quoted in *QTZ*.[48]

Today the term *yingwu*, already found in early sources, is mostly translated as "parrot" or "parakeet", especially if there is no further specification.[49] *Yingwu* birds also figure prominently in different textual genres.[50] In ancient times this

---

46 Descriptions, with references to several subspecies, for example, in *Hainan dao de niao shou*, pp. 231–236.

47 See, for example, *QTZ*, j. 4, 3a–4b. This term also appears in other early sources, beyond the Hainan context – for example, in *(Jiajing) Guangdong tongzhi chugao*, j. 31, 17b.

48 See *Chongbian Qiongtai gao* 重編瓊臺藁, j. 4, 31a–b. For a biography of this man, see *Ming shi*, XVI, j. 181, pp. 4808–4810. A short English account is in Goodrich and Fang, *Dictionary of Ming Biography*, I, pp. 249–252.

49 For *yingwu* in traditional sources, see Schafer, "Parrots in Medieval China"; the same, *The Golden Peaches of Samarkand*, pp. 99–102, and *Vermilion Bird*, pp. 239–240; Ptak, *Exotische Vögel*, pp. 11–33.

50 Currently Katrin Götzinger (Munich) is working on a doctorial dissertation related to this field. For famous "parrot" *fu* 賦, or rhapsodies, see Graham, "Mi Heng's 'Rhapsody on a Parrot'"; Kroll, "Seven Rhapsosides", p. 7; Knechtges, *Wen xuan*, III, pp. 49–56. Generally,

name referred to various species under the *Psittacidae* family, especially to *Psittacula alexandri* (today *feixiong yingwu* 緋胸鸚鵡; in English usually red-breasted parakeet, moustached parakeet, etc.) and *P. derbiana* (*da zixiong yingwu* 大紫胸鸚鵡; Derbyan parakeet, Derby's parakeet, Chinese parakeet, Upper Yangtze parakeet, etc.). In some cases an attribute appears in front of the name. The combinations white / red / green / "five-coloured" *yingwu* are very common, but it is not always clear what they stand for. For instance, "white *yingwu*" normally designates an imported cockatoo, however, in rare cases a different bird may be meant as well.[51]

The *QTZ* gives no colour attribute, we only hear of a variety with a black bill. Moreover, gramatically the sequence "does not change" can refer to the colour of the bill or to the entire bird. In other Hainan chronicles, one often encounters the sequence *lü yi zhu hui* 綠衣朱喙, "green plumage, red bill". These characteristics should point to *P. alexandri*.[52]

Today *P. alexandri*, called *Paleornis alexandri* in earlier zoological works, is the only Hainanese species under the *Psittacidae* family and thus the best candidate for the *yingwu* in *QTZ*. It resembles *P. derbiana* in shape, but is usually smaller than the latter. Both birds have a green tail and very dark or black parts on the forehead, near the eye and behind the bill.

The bills of male *P. alexandri* birds are mostly red, those of the females are mostly black, young birds have entirely black bills. Yet, the traditional name *molai* rarely appears in written sources. Its origin remains obscure. Finally, according to recent zoological works *P. alexandri* is now usually seen on southern and eastern Hainan, not along the west side.[53]

One may add that Robert Swinhoe offers quite a detailed description of this bird (named *Paleornis javanica* in his account). He also says local people would be fond of teaching it to speak. Moreover, "they regard the black-billed and the red-billed birds as distinct. On several occasions I saw them in shops in the town,

---

for a list of bird rhapsodies from Han to Yuan, see Wu Yifeng 吳儀鳳, *Yong wu yu xu shi* 詠物與敘事, pp. 290–298. This book also contains many useful interpretations. For parrots in a religious context: Idema, "The Filial Parrot".

51 See, for example, Ptak, "Weiße Papageien' (*bai yingwu*) in frühen chinesischen Quellen", and "Chinese Bird Imports from Maritime Southeast Asia, c. 1000–1500", especially pp. 230 et seq.

52 See, for example, *(Xianfeng) Wenchang xianzhi* (咸豐) 文昌縣志, j. 2, p. 77; *(Guangxu) Ding'an xianzhi*, j. 1, p. 131; *(Xuantong) Ding'an xianzhi*, j. 1, p. 105; *(Minguo) Gan'en xianzhi*, j. 4, p. 88.

53 On these details see, for example, *Hainan dao de niao shou*, pp. 123–124; Shi Haitao et al., *Hainan luqi beizhui dongwu jiansuo*, p. 133. – For the name *molai*, also see, for example, *(Wanli) Danzhou zhi* (萬曆) 儋州志, tianji, p. 36.

either perched on their triangular frame-cages, to which they were chained, or walking about freely over the counter." Swinhoe did not see many wild *P. alexandri* birds on Hainan, but he encountered them in the Ding'an 定安 region and the Northwest; this could imply a change in the natural habitat from the nineteenth to the second half of the twentieth century.[54]

(7) *Quyu* / Crested Mynas

鴝鵒: 俗呼八哥。取其雛養之，端午剪其舌，能作人言，似鵙而有情。又有牛八哥，稍大而趐白。白八哥堪爲膾。

Translation: It is commonly called *bage* 八哥. One [may] take a small one, raise it and cut [its] tongue on the *duanwu* day 端午; [then] it can imitate human speech, similar to a *ju* 鵙, but more accentuated. There is also the *niu bage* 牛八哥, which is slightly larger, with white wings. The *bai bage* 白八哥 provides delicious meat.

Comment: An alternative form for *quyu* 鴝鵒 is the version *quyu* 鸜鵒 (rarely transcribed *juyu*), from which one can derive the combination *yingyu* 鸚鵒.[55] Normally *quyu* stands for *Acridotheres cristatellus*, the (Chinese) crested myna (also mynah), which is still called *bage* today.

But this bird, classified as *Aethiopsar cristatellus* in ancient European works, is easily confused with *Acridotheres tristis* (*jia bage* 家八哥, the house myna) and *Gracula religiosa* (*liaoge* 鷯哥, also 了哥; grackle, common hill myna), a bird described farther below (see entry no. 8). Both the latter and *A. cristatellus* are good at imitating sounds; this is certainly one reason why many authors were unable to make a clear distinction between them.

The elements *ge* and *bage* occur in several other names, usually with a specification in front of it, as we just saw. There is, for example, the modern name *lin bage* 林八哥, which stands for *A.* (?*fuscus*) *grandis*, the jungle myna, a bird mostly found in Yunnan.[56] Another bird is *A. albocinctus*, now called *bailing bage* 白領八哥 (collared myna). *A. cristatellus* itself sometimes appears as *fengtou*

---

54   Swinhoe, "On the Ornithology of Hainan", pp. 93–94.

55   For details: Roderich Ptak, "Notizen zum *Qinjiliao*, especially p. 451. – Notes in English on the *quyu* are in Schafer, *Vermilion Bird*, p. 244. For personal observation, for example, Hoffmann, "Vogel und Mensch", pp. 48–50.

56   In recent years, zoologists have often made a distinction between *A. fuscus* (*conglin bage* 叢林八哥) and *A. grandis* (*lin bage*). – For an early Chinese depiction of a *lin bage*, see *Gugong niao pu*, I, pp. 50–51.

bage 鳳頭八哥 and even as *liaoge* 了哥 (normally *G. religiosa*, as was just mentioned).

Both *A. cristatellus* and *A. tristis*, as well as *G. religiosa*, are at home on Hainan, but not the other species under the *Acridotheres* group. *A. cristatellus* is found all across the island, *A. tristis* mostly lives in the northern regions, *G. religiosa* belongs to the southern sections, especially the mountainous interior.[57]

The two names *niu bage* and *bai bage* in our text pose questions. *A. tristis* and *A. cristatellus* have some white patches, notably on their wings. Moreover, on average the second bird is slightly larger in size than *A. tristis*. There are also claims that *A. cristatellus* specimen from the Leizhou 雷州 region, opposite of Hainan, would be even larger than the ones collected on the island itself. This suggests some variation within the same species. But whether there is a relation between such variations and the two names in *QTZ*, is hard to tell.

However, perhaps the form *niu bage* is somehow related to yet another term, namely *niubei liao* 牛背鷯.[58] The second character in that combination may have to do with the name *ju* 鵙, also found in the *QTZ* passage, above; this possibility rests on the fact that R154 (read *bei*) in *ju* is phonetically similar to 背. *Ju*, one should add, is similar to *ju* 鶪 and usually thought to stand for shrikes (*bolao* 博勞, various orthographs); it is largely interchangeable with *jue* 鴂, which may also represent these birds, which are listed below (entry no. 27). There are several *bolao* species; these are grouped under the *Laniidae*.

Another and certainly more straightforward explanation for the name *niu bage* comes from later sources, which report that the birds in question fly together in groups and / or often rest on a "cow's back" (*niubei*).[59] In other words, *niu bage* may stem from a combination implying the "*liao* on a cow's back"; alternatively, the last term could be a "free version" based on the earlier (?) form *niu bage*.

A third possibility is to link the form *niu bage* to the name *niushi bage* 牛屎八哥, mentioned farther below, under the entry *baishe* 百舌 (no. 21). In that case we may be looking at *Turdus merula* (*wudong* 烏鶫, common blackbird), which is somewhat similar to the *G. religiosa* (see next item) and / or *A. cristatellus*.

As to the other term, *bai bage*, similar problems prevail. Should one attribute the *bai* colour to the white patches of the *Acridotheres* birds? But why, then, does

---

57   See, for example, *Hainan dao de niao shou*, pp. 201–204; Shi Haitao et al., *Hainan luqi bei-zhui dongwu jiansuo*, pp. 156–157.

58   See, for example, www.cjvlang.com/Birds/starling2.html.

59   See, for example, *(Xianfeng) Wenchang xianzhi*, j. 2, p. 77; *(Minguo) Wenchang xianzhi* (民國) 文昌縣志, j. 1, p. 80; *Yazhou zhi*, j. 4, p. 82. Also the observations in Hoffmann, "Vogel und Mensch", p. 49.

the text say that *niu bage* birds had white wings? Occasionally, in later accounts, one finds the sequence "white under the wings"; would this be more correct?[60]

Furthermore, earlier texts, not explicitly focusing on Hainan, refer to a *bai jiliao* 白吉了. But it is not clear, which bird they mean. One possibility is imported cockatoes. The element *jiliao* also shows up in the name *qinjiliao* 秦吉了, which usually designates *G. religiosa* (see below, entry no. 8).[61] However, later works mention a *baimian liaoge* 白面鷯 (了) 哥, literally a "white-faced *liao-ge*".[62] There is also the name *bai liaoge* 白了哥, which seems to stand for *Sturnus sinensis*, a small bird called *huibei liangniao* 灰背椋鳥 in modern zoology. This is the grey-backed or white-shouldered starling. It comes to Hainan in winter and, indeed, has many grey and some white parts. Other birds under the same family, notably *S. sturnius* (also *Agropsar sturnius*; *bei liangniao* 北椋鳥, Daurian starling) and *S. sericus* (*siguang liangniao* 絲光椋鳥, silky starling), also show grey or white patches. Could one or several of these animals stand for the *bai bage*?[63]

Here we can add one final aspect in regard to the combination *bage*. In other sources – for example *Guangdong xinyu* – we read that one can distinguish *yingwu* birds according to eye colour: *jinliao* 金了 (yellow eyes), 銀了 (white eyes), 鐵了 (black eyes). The text also tells us these birds would be called *liaoge* 了哥, *bage* 唎哥, and *baba* 唎唎. The last combination should be phonetically related to the Portuguese word *papageio*, which in turn is very near to similar expressions in other languages.[64] But most of these phonetic considerations should belong to a later age and are not really relevant in the present context.

The *QTZ* entry mentions the *duanwu* or dragonboat festival, which is celebrated on the fifth day of the fifth lunar month. Other texts contain similar passages, often adding the adverb "slightly" in front of *jian* 剪 ("to cut"), or saying that one should feed the *bage* bird with "wine" and / or hot spices, or teach it to

---

60   See, for example, *(Xuantong) Ding'an xianzhi*, j. 1, p. 105, and *(Guangxu) Ding'an xianzhi*, j. 1, p. 131: both equate *quyu* with *bage*, adding there would also be a *niu bage*, slightly larger in size, with white parts under the wings.

61   Details in Ptak, "Notizen zum *Qinjiliao*", especially pp. 453, 455, 457. Also see entry no. 8, below.

62   See, for example, *(Xuantong) Ding'an xianzhi*, j. 1, p. 105, and *(Guangxu) Ding'an xianzhi*, j. 1, p. 132. – Generally, white birds may have a symbolic significance; see, for example, Soymié, "Le Lou-feou chan", p. 112.

63   See, for example, *Hainan dao de niao shou*, pp. 200–201 (term *bai liaoge* on p. 200).

64   *Guangdong xinyu*, j. 20, pp. 514–515. More in Ptak, "Zhuhai dongwu" 珠海動物, p. 180, and *Exotische Vögel*, p. 17 and note 20 there. For the eyes as a distinctive feature between different types of birds, also *Gugong niao pu*, I, pp. 18 et seq., and other Qing works. This feature goes back to ancient times.

speak on the day in question.[65] The form 剪 is largely interchangeable with 翦, and the expression *jianjian* 翦翦 means something like "eloquent". *Leishu* texts and *bencao* works also mention the tongue-cutting issue, usually by citing earlier material, and there are references to the meat of the *quyu*, said to be "sweet, bland and not poisonous" and thus good against stuttering and other problems.[66]

In conclusion the following may be said: The terms *quyu* / *bage* stand for *A. cristatellus*, *niu bage* and *bai bage* remain unclear, although we suspect that the last term was used for *S. sinensis*. Some later works on Hainan drop the idea of a white variety, but stay with the form *niu bage*. However, the descriptive elements they provide allow no clear identification of this bird.

(8) *Qinjiliao* / Grackles

秦吉了：出西路，俗名了哥，毛色如鴉而小，觜腮俱黃。尤黠慧能言，聲大扵鸚鵡。《外紀》詩并序：唐則天朝，有獻能言鳥者，后喜，爲製鳥名之樂，世傳即此鳥也。

> 口呼萬歲祝千秋，羌入金籠占上流。
> 若到上陽供奉日，隴山鸚鵡更回頭。

Translation: [This bird] comes from the [area along the] "Western Route" and is commonly called *liaoge* 了哥. The feathers are coloured like [the ones of] the crow, but it is smaller, the bill and cheeks are yellow. It is very clever and can speak, with a voice stronger than that of a *yingwu*. A poem and its preface in [*Qiongtai*] *waiji* [read]: When the Tang empress [Wu] Zetian [武]則天 held court, someone presented a bird that could speak; the empress was delighted and had music composed in "its name", which has been transmitted from [one] generation [to the next] and refers to this bird:

> Uttering, the emperor should enjoy one thousand autumns,
> It had to enter a golden cage, [though] placed in a high position.
> On the day it came to Shangyang [Hall], for presentation to [the Empress],
> The *yingwu* from Longshan would think of home.

Comment: The name *liaoge*, literally "brother liao", frequently appears in early sources. For the first character the forms 鷯 and 遼 are common as well. Some

---

65 Examples in Hoffmann, "Vogel und Mensch", pp. 49–50; Ptak, "Zhuhai dongwu", pp. 179–180. Also: *(Minguo) Gan'en xianzhi*, j. 4, p. 90.

66 See, for example, *Chongji Bencao shiyi*, Qin bu, p. 189 (no. 618). Also *Bencao gangmu*, IV, j. 49, p. 2656; Read, *Chinese Materia Medica: Avian Drugs*, pp. 65–66 (no. 296).

texts confuse this bird with the *bage*. The Tang court frequently received exotic birds and, indeed, several stories link these animals to the court of Wu Zetian.[67]

The bird described in this paragraph has the scientific name *Gracula religiosa* (*liaoge*, see previous entry); the most common English names are "grackle" and "common hill myna". In earlier works it is normally placed under the genus *Eulabes*; Swinhoe calls these birds, which he encountered on Hainan, *Eulabes hainanus*.[68]

Until today *G. religiosa* often leads a life in captivity and learns to imitate human speech. In ancient times it was distributed in some sections of continental South China, today its habitat seems mostly restricted to Hainan, although there are reports of ex-captive birds being seen in city parks.[69] Traditional authors have regularly praised the grackle's "linguistic talents", also mentioning its dark colour and yellow waddles under and behind the eyes.

The poem cited from Wang Zuo's work alludes to Empress Wu (624–705), one of the most influential women in imperial China.[70] Wu Zetian was interested in auspicious symbols and exotic creatures, especially birds, which she tried to instrumentalize for political reasons. Court servants trained such birds for performances. The *Tongdian* 通典 even credits the empress with a special piece called "Niaoge wansui yue" 鳥歌萬歲樂, "Birdsong for Ten Thousand Years"; evidently, there were some palace birds which would exclaim "Long live the emperor!" – certainly to the delight and amusement of guests, visitors and some later authors, who tried to make fun of Wu Zetian.[71]

---

67   For details, see Ptak, "Notizen zum *Qinjiliao*", pp. 454–455, and note 21 there.

68   Swinhoe, "On the Ornithology of Hainan", pp. 352–353. Swinhoe also cites the Qiongshan gazetteer, which reports familiar details. This includes differentiation according to eye colour (as in entry no. 7, here) and an unclear reference to a white bird.

69   One recent example: Aomen tebie xingzhengqu, Minzheng zongshu 澳門特別行政區, 民政總署..., *Aomen niaolei* 澳門鳥類, p. 209.

70   The second verse also appears as 恙八金籠佔上流, but this is likely to be wrong. See *Jilei ji*, p. 225. A comment is on: http://www.hi.jcy.gov.cn/danzhou/view.php?xuh=833 (August 2014). – English works on Wu Zetian's rise to power and her rule include, for example, Guisso, *Wu Tse-t'ien and the Politics of Legitimation*; Dien, *Empress Wu Zetian in Fiction and in History*; and Rothschild, *Wu Zhao. China's Only Women Emperor*.

71   See *Tong dian*, IV, j. 146, S. 3721. Also see, for example, *Jiu Tang shu* 舊唐書, IV, j. 29, pp. 1061–1062. – There are several internet sources on Wu Zetian and her birds: See, for example, Norman Harry Rothschild, "Wu Zhao's Remarkably Aviary", *Southeast Review of Asian Studies* 27 (2005), http://www.uky.edu/Centers/Asia/SECAAS/Seras/2005/Rothschild. htm, especially the part on "Speaking Birds"; "Dan'e fu li xun zhenqin" 《儋耳賦》裏尋珍禽: http://www.danzhou.gov.cn:1500/dzgov/dzwh/sanwen/201012/t20101213_534980.html (both August 2014). – Additional references in Ptak, "Notizen zum *Qinjiliao*", n. 21. – Wang

The name Shangyang stands for a building called Shangyang Hall 上陽宮, which belonged to the imperial palace. Wu Zetian's name is linked to that structure. Towards the end of her life, in early 705, she was there and the court undertook efforts to gradually restore the Tang dynasty in her abscence from the main palace buildings.

In early times many authors associated *yingwu* birds, often translated as "parrots", with the Longshan 隴山 area (most likely in Guangxi).[72] All kinds of *yingwu* were kept in the palace, but the prose text tells us, their voices were not as strong as those of the grackle. Put differently, the latter outdid the *yingwu* in imitating human speech. Perhaps this was of disadvantage to the *yingwu*. If so, the last line could be an allusion to opportunistic behaviour; one group of opportunists moved in, others fell in disgrace and disappeared. Or alternatively: *yingwu* birds "turned their head" (for *huitou* 回頭), longing to return home, because they were encaged.[73]

(9) *Wuse que* / Fork-tailed Sunbird

五色雀：蘇軾詩并序：海南有五色雀，常以兩絳者爲長，進止必隨焉，俗謂之鳳凰。云：久旱而見輒雨，潦則反是。吾卜居儋耳城南，嘗一至庭下，今日又見之逸士黎子雲及其弟威家。既去，吾舉酒祝之曰：若爲吾來者，當再集也。已而果然，乃爲賦詩：

粲粲五色羽，炎方鳳之徒。

青黄縞玄服，翼衛兩絨朱。

仁心知閔農，常告雨霽符。

我窮惟四壁，破屋無瞻烏。

惠然此粲者，來集竹與梧。

Zuo himself has an entry called "Niaoge wansui yue" in the collection *Jilei ji*, p. 11. Also see http://www.hnszw.org.cn/data/news/2011/03/48916/ (July 2014) for some of Wang's "bird poems".

72 Scholars disagree on the location of Longshan. Other possible locations include the Gansu / Shaanxi region.

73 *Yingwu* birds often appear in poems with a similar context. See, for example, *Ouyang Xiu quanji* 歐陽修全集, I, j. 4, p. 71. Also see Mi Heng's 禰衡 famous "Yingwu fu" 鸚鵡賦 (references in n. 50, above). Furthermore: *Yuchu xinzhi* 虞初新志, j. 18, pp. 299–300 (two anecdotes about Longshan *yingwu* missing their former owners).

鏘鳴如玉佩，意欲相嬉娛。

寂寞兩黎生[子雲與威]，食菜真臞儒。
小圃散春物，野桃陳雪膚。

舉杯得一笑，見此紅鸞雛。
高情如飛仙，未易握粟呼。

胡爲去復來，眷眷豈属吾？
回翔天壤間，何必懷此都！

《外紀》：五色雀，小鳳凰，

五方色白黑青黃，絳者爲長提其綱。
飛上高枝恰一雙，召呼東西北中央。
四林收聲低翼翔，白黑青黃羅絳傍。
尊卑秩秩如有常，宛若大漢龜茲王。
小小乾坤從翕張，復宜造化知雨暘。
大旱一出雨則滂，久雨纔見陰轉陽。
一羽至末關休祥，況乃應治真鳳凰。

Translation: Su Shi's poem and its preface [read]: On Hainan there are the *wuse que*; they usually consider [a bird] with two crimson wings as [their] leader and definitely follow [his] movements; [the latter] is commonly called *fenghuang* 鳳凰 ("phoenix"). One [also] says, [when the *wuse que*] appears [during] a long drought, sudden rain [will come]; [in the case of] heavy showers, it is the reverse. Once, [when] I had chosen to live in the south of Dan'er 儋耳 city, one [such bird] came down to the courtyard. Today I saw it again in the house of the recluse Li Ziyun 黎子雲 and his brother [Li] Wei [黎]威. As it went off, I raised [my] wine [cup], greeting it: If you come for me, we'll get together again. So it happened and I composed a poem [to record the event]:

Flashing, flaming [its] feathers in five colours:
A follower of the phoenix from the quarter of fire.
Dressed in blue / green and yellow, in white and black,
On [its] wings two ribbons, [all] in red.

Kindhearted, understanding hopeless farmers,
It tells [them] when rain will come and go.

I am poor, with only four walls –
A shattered house, not [even] a crow to look up to.

Gracious [indeed] this flashing [creature]:
It comes [here], perching on bamboo and the parasol tree.
Tinkling its song, like jade pendants,
Wishing to amuse and divert me.

Desolate and lonely the two Li brothers ([text comment]: Ziyun and Wei),
Eating vegetables, truly gaunt scholars.
The small garden, strewn with the creatures of spring,
The wild peach displays [its] snowy skin.

Lifting up a cup, satisfied with a smile,
Seeing this red fabulous young bird.
Feelings elevated [as if I were] a flying immortal,
Not easy to cry out with a handful of millet!

Why does it leave and come again?
[In spite of] affection – could it be close to me?
Returning and hovering between Heaven and Earth,
Why need you cherish this city?

The [*Qiongtai*] *waiji* [says]: The *wuse que*, [or]"little phoenix" [birds],

[Display] the colours of the five quarters: white, black, blue / green, yellow,
The red one being the leader to guide them.
Flying [up to] high branches, [they come] in pairs,
Giving calls to the east, west, north and center,
[While others in] the forest remain silent, lowering [their] wings.
Surrounding the red [bird], the white, black, blue / green and yellow [ones]
Line up [according] to rank, as if having a rule [for that],
Like the Great Han [and] the King of Kucha.
[These birds are] small, [but] Heaven and Earth follow their movements;
Adjusted to Creation and Change, they foreknow rain and sunshine:
[When] they appear [during] a great drought, rich showers [will follow];
Seeing [them] after long-lasting rain: darkness will turn to light.
Even a single feather [of theirs] may indicate blessings,
Certainly this should apply to a true phoenix!

Comment: Before discussing this long entry, the longest in the avian section of *QTZ*, certain "technical" aspects deserve our attention: (1) Schafer published a fine English translation of the first poem (but not of its prose introduction); we have adopted most of his readings.[74] (2) In the modern edition of *QTZ* the first full stop within the prose introduction is placed after 云; this seems wrong. There are also some wrong characters in that edition. (3) For the verses from *Qiongtai waiji* the *QTZ* edition in short characters uses punctuation marks different from the ones chosen here. The status of the initial line starting with "The *wuse que…*" is unclear: it may form part of the poetical structure or one may read it separately. (4) One also finds this second poem in Wang Zuo's *Jilei ji*; there are minor textual variations; these are negligible and of no importance for the translation.[75] (5) In the English version the term *fenghuang* 鳳凰 appears as "phoenix". *Fenghuang* birds, as well as the fabulous *luan* 鸞 creature (verse 18 of Su Shi's poem), are very common in written sources. One also encounters the phoenix on paintings, textiles, ceramics and other objects of arts, which is a complex theme.

The *fenghuang* manifests itself in various ways. According to the *Shanhai jing* 山海經, it resembled "a chicken with the markings of the graph for virtue on its head, duty on its wings, ritual on its back, humaneness on its breast, and trust on its stomach".[76] In the imagination of later readers it was certainly larger than a small *wuse que*, literally "a sparrow with five colours". But the latter's noble behaviour, its habit of resting on parasol or *wu[tong]* 梧[桐] trees (on which the phoenix is said to stay, hence the name "phoenix tree"),[77] the extraordinary colours and the talent to forecast weather changes – all these elements certainly inspired authors to compare the small *wuse* bird to the fantastic *fenghuang*.[78] The texts cited in *QTZ* make this very clear. But the second poem seems different in

---

74 Schafer, *Shore of Pearls*, pp. 106–107. The poem is in *Su Shi quanji*, I, j. 43, p. 534. For an early comment, see Xian Yuqing, "Su Shi yu Hainan dongwu", p. 120.

75 *Jilei ji*, pp. 15–16. The modern notes to this edition are very short.

76 Strassberg, *A Chinese Bestiary*, pp. 102, 193–194, and note 466; *Shanhai jing jiaozhu* 山海經校注, Nanshan, p. 16. There are many more references to the *fenghuang* or *fengniao* 鳳鳥 in that work.

77 *Wu* suggests the combination *wutong*, usually *Firmiana simplex* (*Sterculia platanifolia, Firmiana platanifolia*), less frequently *Aleuritis fordii*. With *tong*, the second character, one can also associate *tonghua* 桐花 (paulownia). Today *Hainan wutong* 海南梧桐 stands for *Firminana hainanensis*. – The names *wu / tonghua / wutong* often appear in literary contexts and there is much confusion in regard to what they mean. See, for example, McCraw, "Along the Wutong Trail: The Paulownia in Chinese Poetry", especially pp. 86–88 (phoenix and *wutong / tong*).

78 See *Taiping yulan*, IV, j. 915, 1a–9b (pp. 4054–4058); *Guangdong xinyu*, j. 20, 509–511; Read, *Chinese Materia Medica: Avian Drugs*, pp. 77–78 (no. 308). – For an early "identification" of the *wucai niao* 五采鳥 ("five-coloured bird", certainly similar to *wuse que*) with *huangniao* 凰鳥, *luanniao* 鸞鳥 and *fengniao*, see *Shanhai jing jiaozhu*, Dahuang xijing, p. 396.

one point: it stresses the ability of *wuse* birds to present themselves in organised groups, with ranks comparable to the difference between the Han emperor and the ruler of the ancient kingdom of Kucha (Junzi / Qiuzi 龜茲).[79] In Su Shi's work the idea of leadership only comes out in the prose introduction.

Above, we had already mentioned that Su Shi spent many years on Hainan. He acquainted himself with local traditions and became familar with the flora and fauna of his new home. The name Dan'er, in the introduction, has a long history; many authors have applied it to the county of Danxian 儋縣 or a larger section of island. Administrative records say a *jun* 郡 (commandry) of that name was established in 110 B.C., under the Han emperor Wudi 武帝.[80] Here, however, it simply evokes the fact that Su Shi's cottage was near the county seat. The Li brothers are also mentioned elsewhere; they lived nearby and were in frequent contact with the poet. "Snowy skin", in the fourth quatrain, alludes to Yang Gui-fei 楊貴妃 (719–756). "The handful of millet refers to the ancient custom of divining the future by giving grain to birds." The word "city" in the last verse is ironic or an expression of hope: the divine bird indicates the secret wish of returning home.[81]

While staying on "Dan'er", Su Shi was not always well-off and certainly thought of moving back to the mainland. This transpires from several of his poems. At the same time, however, he is overwhelmed by the exotic beauty of the *wuse que*, thus combining purely descriptive elements with personal concerns. This is not untypical for Su Shi's art and may be a reason for the fact that later texts frequently cite these lines. Among such works are several geographical accounts which predate the *QTZ*; they list the *wuse que* as one of Hainan's local products (*tuchan* 土產) and refer to Su Shi as well.[82]

Unlike the imagined *fenghuang* the *wuse que* was a real creature. Therefore it also appears in several Guangdong and other Hainan chronicles, many of which cite earlier material, again including Su Shi's poem. Examples are found in Huang Zuo's *Guangdong tongzhi* and the *Guangdong xinyu*; both works carry

---

79   For this polity on the Silk Road, see especially *Han shu*, XII, j. 96b, pp. 3911–3917. – Useful modern works include Liu Mau-tsai, *Kutscha und seine Beziehungen zu China*; Rhie, *Early Buddhist Art of China and Central Asia* (various references, especially in vol. 1); Bielen-stein, *Diplomacy and Trade in the Chinese World, 589–1276*, especially pp. 300–304.

80   See, for example, *QTZ*, j. 2, 2a.

81   See, for example, Jordan, "Su Tung-p'o in Hainan", especially pp. 34–35; Hargett, "Clearing the Apertures and Getting in Tune", pp. 153–156; Schafer, *Shore of Pearls*, p. 107, for "snowy skin" and the rest.

82   For example: *Yudi jisheng* (1221), IV, j. 124, 9a (p. 3569); *Fangyu shenglan* 方輿勝覽 (1239), II, j. 43, p. 771; *Da Ming yitong zhi* 大明一統志, IX, j. 82, 22a..

long chapters on that bird.[83] These and other texts say the *wuse que* was a phoe-nix from the Luofu mountains 羅浮山, a sacred place in Bolou county 博羅縣 of Guangdong; this is why our bird also bears the name *Luofu feng* 羅浮鳳, literally the "Luofu phoenix".[84]

Like all other famous mountains in China, the Luofu is surrounded by fan-tastic stories.[85] But instead of discussing this material, we should look at further details pertaining to the *wuse que*. From local gazetteers we learn the following: The Hainanese *wuse que* is closely related to, yet different from the Luofu va-riety. The latter is normally smaller than a "parrot" (or *yingwu*) and has either two or five colours. Usually a red bird leads the group and the ones with mixed colours are his "subjects". Occasionally black birds with an "iron" comb also appear as leaders. By contrast Hainan's *wuse que* have five colours and there are no special references to a black kind. Moreover, those with five colours predict weather changes, while the Luofu birds come out of the forests to greet noble visitors.[86]

This last feature explains another name: *muke niao* 木客鳥. According to the fragmentary *Yiwu zhi* 異物志, quoted in *Taiping yulan* and many other works, a mountain deity sends such birds to welcome visitors. The *muke niao* are as large as magpies (*que* 鵲) and appear in crowds. But the distribution of colours and ranks among them differs from the above.[87] Besides the brief reference to the *muke niao*, the *Taiping yulan* also provides entries on *shenque* 神雀, *chique* 赤雀, *baique* 白雀, *huangque* 黃雀, *qingque* 青雀 and *daque* 大雀, all with citations from earlier works. Interestingly in several cases the *shenque* is said to have five colours as well. But the identification of these *que* birds remains difficult.[88]

---

83   *(Jiajing) Guangdong tongzhi*, II, j. 24, 12b–14a (pp. 630–631), and *Guangdong xinyu*, j. 20, pp. 515–516. Some characters in Su Shi's poem vary from citation to citation.

84   See, for example, *(Jiajing) Guangdong tongzhi chugao*, j. 31, 18a; *Luofushan zhi huibian* 羅浮山志會編, j. 7, 1a–b (p. 182). – For a discussion, see Wang Ting 王頲, "Feng sou li yu" 鳳藪麗羽, especially p. 119. Also Mathias Röder, "Vom kopfüber Hängenden oder *daoguaniao*", especially pp. 20, 23.

85   Soymié, "Le Lou-feou chan", presents an exhaustive analysis of the *Luofushan zhi huibian*.

86   See the detailed entry in *Guangdong xinyu*, j. 20, pp. 515–516.

87   *Taiping yulan*, IV, j. 927, 4a–b (p. 4120). Also see, for example, *(Jiajing) Guangdong tongzhi*, II, j. 24, 13a (p. 631); *Guangdong xinyu*, j. 20, p. 515. Furthermore: Soymié, "Le Lou-feou chan", pp. 45–46 and p. 46 n. 1, and Xian Yuqing, "Su Shi yu Hainan dongwu", p. 120. – Evidently, the comparison with magpies has also led to the name *wuse que* 五色鵲, for which there is a separate entry in *Sancai zaoyi*. At the same time this work offers a sec-ond entry for the familiar *wuse que* and a third one for *muke niao*; see j. 24, 18a–b (p. 481), and j. 24, 9a (p. 390).

88   *Taiping yulan*, IV, j. 922, 5b–12a (pp. 4092–4095).

There are many more names for the *wuse que* and / or the Luofu birds. They mostly appear in local chronicles and are sometimes associated with certain forms and colours. But these names pose further questions or relate to totally different animals. Only one name, *hedan* 鶡鴠 (also *handan* 鳱鴠), shall be mentioned here. According to some sources this animal has green feathers and a red bill. In other cases one finds very different descriptive elements.[89]

Can we identify the *wuse que*? – The following observations may still be of relevance: While Su Shi speaks of multi-coloured birds with "ribbons" or stripes on the wings, the second poem and certain other sources suggest the existence of different birds, each with one major colour, the full ensemble of which should correspond to the heavenly directions. Secondly, most texts say the *wuse que* were small; only in exceptional cases are they described as large creatures.[90] Finally, they live in groups. But one can also read Su Shi's poem differently: *wuse que* birds are rare, therefore the author feels happy to have seen one or two specimen of these beautiful creatures.

If the image is one of multi-coloured small birds, we may be looking at a member of the *Nectariniidae* family. Today most of these birds are endangered. *Aethopyga gouldiae* (*lanhou taiyangniao* 藍喉太陽鳥; blue-throated sunbird, Gould's sunbird) is still found in Guangxi, but not in Guangdong. *A. gouldiae* has several subspecies and moves around in small flocks. Another member of the *Nectariniidae*, *Aethopyga christinae* (*chawei taiyangniao* 叉尾太陽鳥, also *yanwei taiyang niao* 燕尾太陽鳥; fork-tailed sunbird), has less spectacular colours; one finds it in Guangdong and on Hainan; however, it rarely seems to appear in large groups (?). Many works refer to a Hainanese subspecies of the latter, called *A. c. christinae*.[91] The conclusion is that Su Shi could mean this bird, while the second poem contains elements that should better suit the first candidate.

---

89 Different names: *Guangdong xinyu*, j. 20, 515–516. Further suggestions, in connection with the so-called *yinsheng niao* 音聲鳥, in Sun Shu'an 孫書安, *Zhongguo bowu bieming da cidian* 中國博物別名大辭典, p. 507. – Examples in sources: *(Jiajing) Guangdong tongzhi*, II, j. 24, especially 12b (p. 630); *Luofushan zhi huibian*, j. 7, p. 182. Soymié defines the *wuse que* with green feathers and a red bill as "un espèce des perroquets"; see"Le Lou-feou chan", p. 46 n. 1. – The *hedan* is also called *hanhao chong* 寒號虫 (sometimes: flying fox), which links to a very different theme; see, for example, *Bencao gangmu*, IV, j. 48, pp. 2642 et seq.; Read, *Chinese Materia Medica: Avian Drugs*, pp. 57–59 (no. 290).

90 One case is in *Yazhou zhi*, j. 4, pp. 81–82: the bird's tail is four to five *chi* 尺 (feet !) in length, the colour pattern is similar to that of a peacock.

91 See, for example, *Hainan dao de niaolei*, pp. 270–271. An early description is in Swinhoe, "On the Ornithology of Hainan", pp. 236–237.

The association of sunbirds with the term *wuse que* was also suggested by other authors.[92] Some of them discuss further names, for example, various combinations with the character *tong* 桐 (the second graph in *wutong*). This includes such compounds as (*tonghua*) *wuse lingqin* (桐花) 五色靈禽, *tonghua feng* 桐花鳳 and *tonghua niao* 桐花鳥; in these cases *tonghua* evokes paulownia flowers. Su Shi himself mentions such birds. But this takes us to different terminological horizons involving such expressions as *daoguaniao* 倒挂鳥 and thus to a different ornithological terrain.[93]

(10) *Zhiji* / Pheasants and Others

雉鶏: 其屬有錦葉、金錢之類。

Translation: Among its kind are the [ones] with a *jinye* 錦葉 (golden-leaf) and a *jinqian* 金錢 (golden-coin) [pattern].

Comment: Classical Chinese sources refer to different kinds of *zhi* 雉, a term usually translated as "pheasant" in English. But the English word may not always be appropriate because it evokes a variety of elegantly looking, long-tailed birds such as *Chrysolophus pictus* (*hongfu jinji* 紅腹錦鶏, golden pheasant) and *Phasianus colchicus* (*huanjing zhi* 環頸雉 or *zhiji*, as the name of this entry; common pheasant, ring-necked pheasant) both of which are not at home on Hainan,[94] while the expression *zhi* by itself, or in combination with other characters, may also refer to various short-tailed animals with a less spectacular colour pattern.

Zhi birds appear in many early Chinese poems and other texts. Works in the tradition of the *Erya* 爾雅 and *Shuowen jiezi* 說文解字 provide long lists of related names.[95] Additional terms are found, for example, in local gazetteers. This

---

92 See, for example, Li Haixia 李海霞, "Daxing cidian niao shou citiao shiyi jiubu" 大型詞典鳥獸詞條釋義糾補, p. 63 (pp. 10–11 of electronic version: http://www.docin.com/p-729813761.html; August 2014); Wu Jiayi 吳佳翼, "Nanfang xiao fenghuang – wuse que" 南方小鳳凰 – 五色雀, pp. 99–103.

93 For details in modern studies: Röder, "Vom kopfüber Hängenden oder *daoguaniao*", pp. 20–22, and Wang Ting, "Feng sou li yu", part 4. – A traditional source with many references: *Guangdong xinyu*, j. 20, pp. 509–511. – A very early European reference to the "Tung-hoa-feng" in Du Halde, *Description*, I, p. 33.

94 One or two works state that *P. colchius* would be on Hainan (for example, Zhongguo ye-sheng... Qian Yanwen, *Zhongguo niaolei tujian*, p. 96), but this should be wrong.

95 See, for example, *Erya zhushu* 爾雅註疏, j. 10, pp. 317–318 and *Shuowen jiezi zhu* 說文解字注, si pian, shang, 25b–26a (pp. 141–142). – Modern accounts with useful notes on pheasants in traditional China include, for example, Schafer, *Vermilion Bird*, pp. 243–244, and

includes such expressions as *yeji* 野雞,[96] *huachong* 花虫,[97] *shanji* 山雞, or *junyi* 鵔鸃, *zhidi* 雉翟,[98] etc. Some of these names were also used for various other birds. *Leishu* works such as the *Taiping yulan* give access to further details.[99]

Clearly, the large number of names makes it difficult to identify species. Moreover, the borderline between *zhi* 雉 birds and *ji* 雞 birds (usually "chicken") seems quite flexible.[100] The term *yeji* is a good example: normally it stands for the precursor of the house chicken, but it also comes up in the context of "pheasants". Later entries on *zhi* birds in Hainanese chronicles say the term *yeji* was used for both the *jinhua* 金華 (same as *jinye*) and *jinqian* varieties, or just for the latter, depending on one's reading.[101]

The name *shanji* is another case full of ambiguity. It appears farther below (see entry no. 39). This animal has "white ears", as we shall see; moreover, later texts, quoting earlier material, say it would dance and love its feathers, besides being good at fighting.[102]

Today, Hainan has seven species that belong to the *Phasianidae* family. The attribute *jinqian*, or "golden coin" (which is also used to describe the patterned fur of certain mammalia), clearly suggests *Polyplectron bicalcaratum* (or *P. katsumatae*), i.e., the (Hainan) grey peacock pheasant. Chinese zoologists call this bird *hui kongque zhi* 灰孔雀雉 or *kongque ji* 孔雀雞, but it is still known under its popular name *jinqian ji* 金錢雞 (which also occurs in traditional texts). Its plumage shows a dotted coin-like pattern and carries a metallic lustre, while its tail feathers are shorter than those of, say, *P. colchicus*. The "golden-leaved" variety (*jinye / jinhua*) is less easy to identify. The other Hainanese members of the *Phasianidae* are distinctly different in shape and there is no bird that one can readily associate with the *jinye* kind.[103]

Guo Fu, *Zhongguo gudai dongwuxue shi*, pp. 431–434. – A general zoological work in English is Li Xiangtao 李湘濤, *Gamebirds of China*.

96 See, for example, *(Qianlong) Qiongzhou fuzhi*, j. 1 xia, 97b (p. 103); *(Kangxi) Qiongshan xianzhi* (康熙) 瓊山縣志, j. 9, 25a (p. 542).

97 See, for example, *(Guangxu) Ding'an xianzhi*, j. 1, p. 130.

98 For the last three especially *(Jiajing) Guangdong tongzhi*, II, j. 24, 10b (p. 629), 14a (p. 631).

99 *Taiping yulan*, IV, j. 917, 4b–9a (pp. 4066–4069). Also see, for example, Read, *Chinese Materia Medica: Avian Drugs*, pp. 38–41, and sources there.

100 References in *(Jiajing) Guangdong tongzhi* suggest this, to mention just one case.

101 See *(Kangxi) Qiongshan xianzhi*, j. 9, 25a (p. 542); *(Qianlong) Qiongzhou fuzhi*, j. 1 xia, 97b (p. 103); *(Daoguang) Qiongzhou fuzhi* (道光) 瓊州府志, j. 5, 45a (p. 141 top).

102 See, for example, *(Guangxu) Chengmai xianzhi* (光緒) 澄邁縣志, j. 1, p. 81; *(Qianlong) Qiongzhou fuzhi*, j. 1 xia, 95b (p. 102); *(Minguo) Danxian zhi*, j. 3, p. 188.

103 For this family on Hainan: *Hainan dao de niao shou*, pp. 66–74; Shi Haitao et al., *Hainan luqi beizhui dongwu jiansuo*, pp. 117–118. – For *P. bicalcaratum*, for example, Li Xiangtao,

A further examination of traditional sources leads to very little, or even complicates the issue. Dai Jing's version of the *Guangdong tongzhi* is a case in point. This text lists several *ji*, among which one finds a *jinqian ji* and a *zhiji*. Of the latter it says that it had a "*jinhua* tail and a *jinqian* pattern" (金華尾金錢紋). In other words, the two expressions used in *QTZ* to describe different kinds, here appear as attributes associated with one and the same bird.[104] However, if we define this bird as *P. bicalcaratum*, then what does the term *jinqian ji* stand for? There also arises another problem: *P. bicalcaratum* is not at home in what is now Guangdong; perhaps this was already so under the Ming. The conclusion would then be that on the continent the names in question should apply to different birds.

There is no solution to this terminological dilemma. It is possible that in earlier times Hainan still had one or two additional kinds of "pheasants", which have disappeared in the course of time. One may also assume that the terminology, introduced from the Chinese mainland, underwent local modifications. Nevertheless, as was said, the *jinqian* variety of the *zhiji* should represent *P. bicalcaratum*.

(11) *Baixian* / Silver Pheasants

白鷴: 冠距觜丹，白質黑章。

Translation: Its comb is huge, its bill red, and its body white with black ripples.

Comment: The *baixian*, literally "white *xian*", has a long history and appears in many early texts. Interestingly, Ge Hong 葛洪 (283–343; varying dates) mentions it together with a *heixian* 黑鷴, or "black *xian*", both of which were offered to the Han court from Nan Yue 南越.[105]

Gamebirds of China, pp. 143–145. – For the name *jinqian ji* in traditional texts, see, for example, *(Daoguang) Guangdong tongzhi*, III, j. 99, p. 248 bottom.

104 *(Jiajing) Guangdong tongzhi chugao*, j. 31, 18a.

105 *Xijing zaji* 西京雜記, j. 4, p. 25. Translation in Heeren-Diekhoff, *Das* Hsi-Ching-Tsa-Chi. *Vermischte Aufzeichnungen über die westliche Hauptstadt*, p. 183. For an English work, see Nienhauser, *An Interpretation of the Literary and Historical Aspects of the* Hsi-ching tsa-chi *(Miscellanies of the Western Capital)*. Note: the authorship of *Xijing zaji* is disputed. – The statement found in *Xijing zaji* is often repeated in later sources; see, for example, *(Jiajing) Guangdong tongzhi*, II, j. 24, 14b (p. 631). – For a detailed study of the *baixian* in traditional sources, see Wang Ting 王頲, "Shu shi qiu zhi – Tang, Song liang dai baixian xuyang ji qixi di kao" 蜀士求雉 – 唐、宋兩代白鷴畜養及棲息地考, which is available online under http://www.historicalchina.net/admin/WebEdit/UploadFile/WhiteWT.pdf (June 2014). For short notes on its domestication: Guo Fu, *Zhongguo gudai dongwuxue shi*, pp. 433–434.

Evidently the white *xian* bird became a symbol of good fortune. Under the Ming and Qing dynasties its picture also appeared on official court robes associated with the fifth rank.[106] Such a choice was not accidental: *xian* birds had a reputation of being elegant, males and females were thought to be loyal to each other and considered well-mannered.[107] Li Bai 李白 (701–762) praises these birds for their beauty and gentle character; many other writers also describe them in their works.[108] Of course, the male animals are not as peaceful as one may think: when kept in cages, they begin to fight and must be separated. But in former times the rich raised them in parks and gardens.[109]

One long entry on this bird can be found in Qu Dajun's account which summarizes all essential features associated with this exotic creature that also had the poetic name *jianke* (or *xianke*) 閒客, "idle guest". A further entry, partly with similar details, is contained in the Daoguang version of *Guangdong tongzhi*, to mention just one additional example.[110]

The identification of the *baixian* seems to pose no problem. This is the so-called silver pheasant, or *Lophura nycthemera*, under the *Phasianidae* family; its modern name still is *baixian*. However, there are several subspecies with slightly different colour patterns. These belong to the avifauna of Fujian, Guangdong and other parts of southern China. The variety found on Hainan is called *L. n. white-headi*. The male bird of that kind has a white back and white wings with a distinct v-shaped decoration (this seems to echo the term *zhang* 章 in *QTZ*), while its crest and breast are black, sometimes with a bluish shine. It also has a long tail. The female bird has many grey and brown parts, also with a dotted pattern.[111]

Yet, there are two elements in *QTZ* – and other chronicles with near-to-identical wording – that raise questions: the attribute "red" and the term *ju* 距 behind

---

106 See, for example, Zhang Rong 張榮, "Yi fang Qingchao wenguan buzi de shenshi" 一方清朝文官補子的身世.

107 For example, *(Qianlong) Qiongzhou fuzhi*, j. 1 xia, 96a (p. 102), and *Guangdong xinyu*, j. 20, pp. 513–514; Read, *Chinese Materia Medica: Avian Drugs*, p. 42 (no. 273).

108 See *Li Bai ji jiaozhu* 李白集校注, II, j. 12, pp. 808–809: "Zeng Huangshan Hu gong qiu baixian" 贈黃山胡公求白鷴 (dated 754). Li Bai's poem has attracted much attention and there are many internet entries on it. Also see, for example, Schafer, *Vermilion Bird*, p. 244. – A later example of verses on this bird is in *Wanling ji* 宛陵集, j. 4, 11a.

109 See Guo Fu, *Zhongguo gudai dongwuxue shi*, pp. 433–444.

110 See *Guangdong xinyu*, j. 20, p. 513; *(Daoguang) Guangdong tongzhi*, III, j. 99, p. 248 bottom.

111 See *Hainan dao de niao shou*, pp. 70–71; Shi Haitao et al., *Hainan luqi beizhui dongwu jiansuo*, p. 118. – Li Xiangtao, *Gamebirds of China*, pp. 116–119, provides a general account in English.

the word "crest".[112] *Ju* can also mean "spurs". The feet of *L. nycthemera* are red, but not the bill; the red colour should refer to the fleshy parts around the eye. These observations also apply to the Hainanese kind.

(12) *Yao* / Sparrow-hawks or Harriers

鷂: 俗名鷂鷹。

Translation: Its common name is *yaoying* 鷂鷹.

Comment: Old works often equate the name *yao* with other names, for example, *queying* 雀鷹 and *yaozi* 鷂子. *Yao*, according to the *Shuowen jiezi*, stands for a bird of prey.[113] In English it often gets translated as "sparrow-hawk", "harrier", "kite", etc. Such birds feed upon small animals, including sparrows. This also incited Cao Zhi 曹植 (192–232), a famous writer of the late Han and early Sanguo period, to create a fable in which a sparrow outwits a hungry *yao* and thus avoids being devoured by the latter.[114]

Besides moving into literary terrain, *yao* birds have entered proverbs and appear in the context of hunting – similar to falcons, eagles, etc. Indeed, there is much written and pictorial evidence of "falconry" in the Far East, but this usually excludes southern China and Hainan, although there are exceptions to the rule.[115]

The character *ying* 鷹 (in *yaoying*) also occurs below (see entries no. 14 and n. 52). Usually it represents a hawk. But this of little help in trying to identify the meaning of the compound *yaoying* – or the single term *yao*. The *QTZ* provides

---

112 For example, *(Jiajing) Guangdong tongzhi chugao*, j. 31, 18a.

113 *Shuowen jiezi zhu*, si pian, shang, 51a (p. 154). Also see, for example, Ptak, "Zhuhai dong-wu", p. 176 (no. 15). – Generally, for birds of prey in China and their classification, see, for example Li Xiangtao, *Raptors of China*.

114 See *Cao Zhi ji jiaozhu* 曹植集校注, j. 2, pp. 302–305: "Yao que fu" 鷂雀賦. Also see, for example, Wu Yifeng, *Yong wu yu xu shi*, pp. 90–93. There are many European-language works on Cao Zhi not listed here.

115 Standard works are Schafer, "Falconry in T'ang Times"; Wallace, "Representations of Falconry in Eastern Han China (A.D. 25–220)". – Schafer's article contains the translation of an early text, the *Roujue bu* 肉攫部, attributed to Duan Chengshi, and included in the latter's *Youyang zazu*, j. 20. For this work also see, for example, Guo Fu, *Zhongguo gudai dongwuxue shi*, p. 441; Reed, *A Tang Miscellany. An Introduction to* Youyang zazu (New York, etc.: Peter Lang, 2003), especially p. 69. – For the *Youyang zazu*, also Schafer, "Notes on Duan Chengshi and His Writing". More on birds of prey in his *Golden Peaches*, pp. 93–96. – For brief bibliographical information on *Roujue bu*, see Siebert, *Pulu*, pp. 245 and n. 447, pp. 291, 298.

no information on the size and colour of this bird, so we can only turn to other sources for possible explanations. One early observer is Ge Hong. He refers to different *ying* birds and two kinds of *yao*: *congfeng yao* 從風鷂 (wind followers) and *gufei yao* 孤飛鷂 (lonely flyers).[116] Further animals associated with the *yao* "class" include the following birds: *tijian* 鶙鵳, *yuzi* 鷸子, and *longtuo* 籠脫. *Tijian*, usually defined as a *sun* 鶽 (generally: falcon), is good at catching sparrows. The size of the *yuzi* is like that of a *hu*-swallow (we had encoured this name in entry no. 1), but it also eats sparrows. The *longtuo* differs in that regard: it is strong enough to attack *jiu* 鳩 (various explanations, for example, turtle doves) and *que* 鵲 (magpies).[117]

Li Shizhen, to quote another traditional source, provides no separate entry on the *yao*, but he mentions that *yao* birds are smaller than hawks (*ying*) and have rudder-like tails. He also records their excellent vision, their ability to fly high and to descend with enormous speed.[118]

Ornithologists have spotted many different birds of prey on Hainan. Swinhoe lists various kinds, trying to identify them, but his observations are not always helpful.[119] Today these birds are grouped under the *Accipitridae* family. Several of them carry the character *yao* in their current scientific names and belong to the genus *Circus*: *C. cyaneus* (*baiwei yao* 白尾鷂; white-tailed sparrow-hawk, marsh hawk or hen harrier), *C. macrourus* (*caoyuan yao* 草原鷂; grassland sparrow-hawk, steppe or pale harrier), *C. melanoleucus* (*queyao* 鵲鷂; sparrow-hawk or pied harrier), *C. aeruginosus* (*baitou yao* 白頭鷂; while-headed sparrow-hawk, swamp hawk or western marsh harrier), and *C. spilonotus* (*baifu yao* 白腹鷂; white-bellied sparrow-hawk or eastern marsh harrier). One may add that the taxonomy has undergone many changes and that one finds contradictory details in the literature, especially in regard to the geographical distribution of certain species.[120]

The *QTZ* gives no details regarding the colour, size and general appearance of the *yaoying*. Therefore the term *yaoying* could refer to one or several of the birds listed above, especially to the smaller varieties.

---

116 *Xijing zaji*, j. 4, p. 30; Guo Fu, *Zhongguo gudai dongwuxue shi*, p. 439. – For a recent discussion Mühlbauer, *Greifvögel und Beizjagd in Asien*, pp. 40–43.

117 See, for example, *Taiping yulan*, IV, j. 926, 7b–9a (pp. 4116–4117). Also see *Guangya shuzheng* 廣雅疏証, j. 10 xia, p. 377. The roots of this work go back to the early 3rd century.

118 *Bencao gangmu*, IV, j. 49, p. 2671; Read, *Chinese Materia Medica: Avian Drugs*, p. 85.

119 Swinhoe, "On the Ornithology of Hainan", pp. 84 et seq.

120 See *Hainan dao de niao shou*, pp. 61–63; Shi Haitao et al., *Hainan luqi beizhui dongwu jiansuo*, p. 114.

(13) *Wu* / Crows or Ravens

烏：能反哺，其頸白者名鴉。

Translation: It can recycle food [to its parents]. The ones with a white neck are called *ya* 鴉.

Comment: The word *wu* 烏 means "black" and evokes the image of a dark creature whose body parts, such as the eyes, are not easily discernible.[121] This may explain why the *wu* sometimes appears as a *xuanniao* 玄鳥, or "gloomy bird", in classical sources.[122] Darkness tends to be associated with serious events, thus the *wu* has gradually become a symbol of misfortune. Its negative connotation is implicitly present in several sayings, for example, in the phrase *tianxia wuya yiban hei* 天下烏鴉一般黑, i.e., "all crows under Heaven are black". Furthermore, the *wu* bird's voice sounds *ya ya* – an inauspicious sound no one likes to hear. Therefore, someone who is very talkative, or gives others an unpleasant feeling, is depicted as having a *wuya zui* 烏鴉嘴, or "crow's mouth".

However, "black" birds were also instrumentalised by fortune tellers and they were not always seen negatively.[123] The *Taiping yulan* contains a long chapter with many references to old texts which tell us how *wu* birds were perceived in different periods.[124] Here one enters the domain of mythology. This includes a *wu* with three legs and the *wu's* connection to the sun.[125]

Another theme is the *wu's* association with filial behaviour, expressed in the combination *xiaoniao* 孝鳥, or "the filial bird".[126] The attribute "filial", described

---

121 See Han Xuehong 韓學宏 et al., *Jingdian Tang shi niaolei tujian* 經典唐詩鳥類圖鑒, p. 66. For general information, see pp. 64–67 there.

122 See, for example, *Gujin zhu* 古今注, j. 2, p. 13. Normally *xuanniao* stands for other birds, especially swallows (see entry no. 1).

123 See, for example, *Bencao gangmu*, IV, j. 49, p. 2661; Read, *Chinese Materia Medica: Avian Drugs*, pp. 70–71 (no. 302); references in Sterckx, *The Animal and the Daemon*, pp. 295–296 n. 177. In ancient times there was a *Ya jing* 鴉經, the *Classic of Ravens*, certainly a book on the art of divination.

124 *Taiping yulan*, IV, j. 920, 1a–8a (pp. 4081–4084). Other collections also carry long entries on these birds, for example, *Piya*, II, j. 6, 10b–12a.

125 See, for example, Yang and An, *Handbook of Chinese Mythology*, pp. 95–96; Strassberg, *A Chinese Bestiary*, pp. 208–209; Allan, *The Shape of the Turtle*, pp. 30–33; Sterckx, *The Animal and the Daemon*, p. 63; *Shanhai jing jiaozhu*, Dahuang dongjing, p. 354 and p. 355 n. 7; *Huainanzi jishi* 淮南子集釋, II, j. 7, pp. 508–509;

126 *Shuowen jiezi zhu*, si pian, shang, 4a, 56a (p. 157); *Gujin zhu*, j. 2, p. 13, and many later works.

in many sources, derives from the bird's ability to regurgitate food and thus to feed its own parents.[127] This also explains the expression *fanbu* 反哺 in our text.

Evidently, the "filial" properties of the *wu* were a fascinating subject, which caught the attention of many scholars. One author, Cheng Gongsui 成公綏 (227–271), tells us he once saw a *xiaoniao* visiting his house whereupon he exclaimed: "I do not have the virtue of benevolence! Why have these auspicious fowl come to visit me?" He then goes on praising the *wu* for its filial qualities.[128]

Wu birds did not only symbolize filial behaviour, to some they became a sign of welfare and prosperity. Such birds would appear when a country was ruled by righteousness; they would not be seen in locations without virtue.[129]

But this is not the whole story: Local gazetteers and sources of later periods declare that some *wu* were not filial at all. The black ones, which feed their parents, are called *ciwu* 慈烏 (literally: kind crow") or *ciniao* 慈鳥 (kind bird); by contrast, the ones with a white neck do not provide such extraordinary services and are named *ya* 鴉, or 鵶.[130] We also hear of yet another type of *wu*; this is the *huoya* 火鴉 (literally: fire-crow), which devours flames. The *huoya*, so it seems, is mostly associated with Danzhou 儋州; it likes to kindle a fire on the roof, while using its wings to fan the flames. To avoid such disasters, people would offer food sacrifices to that bird.[131]

In the notes above the terms *wu* and *ya* were already rendered as "crow". This takes us to the *Corvidae* family. Modern zoology lists four kinds of birds under this family that are found on Hainan: *Corvus frugilegus* (*tubi wuya* 禿鼻烏

---

127 See, for example, Read, *Chinese Materia Medica: Avian Drugs*, p. 69 (nos. 301, 301A, 301B); *Bencao gangmu*, IV, j. 49, p. 2660–2662.

128 See the preface to his "Wu fu" 烏賦. An analysis of this text is in Wu Yifeng, *Yong wu yu xu shi*, pp. 106–107.

129 *Taiping yulan*, IV, j. 920, 4a (p. 4082), quoting Cheng Gongsui, *Shuo yuan* 說苑, and other works. For the *Shuo yuan*, see, for example, Stumpfeldt, *Ein Garten der Sprüche*, II, p. 393. – *Wu* birds also entered later poetry, where they were used in various ways, as symbols of bad luck, but also in other contexts. One famous text is in *Wanling ji*, j. 60, 4a–5a: Mei Yaochen's 梅堯臣 "Lingwu fu" 靈烏賦. For an interesting study on *wu* uses, see, for example, Tan Mei Ah, "Beyond the Horizon of an Avian Fable", passim.

130 See, for example, *(Qianlong) Lingshui xianzhi* (乾隆) 陵水縣志, j. 1, pp. 133–134; *(Xianfeng) Wenchang xianzhi*, j. 2, p. 76; *(Guangxu) Ding'an xianzhi*, j. 1, p. 130; *(Minguo) Gan'en xianzhi*, j. 4, p. 89; *Yazhou zhi*, j. 4, p. 81. – The *(Kangxi) Lin'gao xianzhi* (康熙) 臨高縣志, j. 2, pp. 52–53, lists a "kind" and a *ya* bird separately.

131 See, for example, *(Minguo) Danxian zhi*, j. 3, p. 189; *Guangdong xinyu*, j. 20, p. 526. – *Huoya* also appear in magic warfare, for example, in the novel *Nanyou ji* 南遊記, hui 9. See Ptak, "Qianliyan und Shunfeng'er in Erzählungen und anderen Texten der Yuan- und Ming-Zeit", to be published in a *Festschrift* for P. Roman Malek.

鴉, rook), *C. corone* (*xiaozui wuya* 小嘴烏鴉, carrion-crow), *C. macrorhynchus* (*dazui wuya* 大嘴烏鴉, large-billed or thick-billed crow) and *C. torquatus* (*baijing ya* 白頸鴉, collared crow). The status of the first two is not very clear, but the other two have been recorded in many locations. All these birds are mostly black, only *C. torquatus* has some white parts, notably around its neck, as the name suggests. This should be the most likely candidate for the "unfilial" variety, while the other type might then be *C. macrorhynchus*. In general, however, all four could match the *wu* / *ya* in *QTZ*.[132]

(14) *Ying* / Hawks or Eagles

鷹：蒼黑色。又種倉褐色，似鴟。

Translation: It is dark [or] black in color. Another kind is dark brown and looks like a *chi* 鴟.

Comment: Usually the term *ying* stands for a medium or large-sized bird of prey. Such birds have entered the mythology and literature of many cultures; they are praised for their strength, fierce character, excellent vision, flying skills and high speed. Ancient China is no exception in that regard. Poems, lexicographic and other works are full of references to *ying* birds.

One long entry is found in the *Bencao gangmu*, which refers to Wei Yanshen 魏彥深 of the Sui dynasty, whose "Ying fu" 鷹賦 is well-known. This is what Read has to say: "The claws of the best kind are shaped like a + character, the best kind of tail is the black type in which the feathers are united. The beak is hooked, the legs are like dried sticks. The plumage is either a kind of tabby white, or glossy like varnish. The stripes are like embroidery and a fine netted line. It is as heavy as a mass of metal with claws as strong as steel. The plumage is constantly changing colour until it assumes a general yellow colour. At two years it is called a Yao, at 3 it is a 蒼 Ts'ang. The female is larger than the male. If eaten one must be careful to take ginger and wine, the former to expel the heat the latter to drive out the cold. Those born in mountain caves are quiet, but those nesting in trees are always on the alert. The foot being long it rises slowly, the six quills in either wing are short hence they move rapidly."[133]

---

132 See *Hainan dao de niao shou*, pp. 210–212; Shi Haitao et al., *Hainan luqi beizhui dongwu jiansuo*, p. 158. – Swinhoe, "On the Ornithology of Hainan", pp. 348–350, already collected valuable information on these birds.

133 Read, *Chinese Materia Medica: Avian Drugs*, pp. 81–82. – For the poem see, for example, *Bencao gangmu*, III, j. 49, pp. 1759–1760; *Taiping yulan*, IV, j. 926, 5a–5b (p. 4115); Guo Fu, *Zhongguo gudai dongwuxue shi*, pp. 440–441.

Although these lines give interesting details, species identification remains difficult. Traditional works offer no systematic coverage of possible differentiae (colour, types of prey, eyes, the shape of wings, etc.). The *Taiping yulan* and similar works contain many references to *ying* birds, among which one finds, for example, a *cangying* 蒼鷹 ("dark", "blue" or "grey" *ying*, depending on the interpretation of *cang*), a *zhiying* 雉鷹 ("pheasant hawk"), a *tuying* 菟鷹 ("rabbit hawk"), a *qingchi* 青翅 ("blue / green-" or "black-winged [hawk]"), a *caomou* 草眸 ("green-eyed [hawk]"), a *qingming* 青冥 ("blue / green" or "black [hawk]") and a *jinju* 金距 ("golden-claw [hawk]"); however, defining these birds in terms of modern taxonomy is nearly impossible.[134]

Like the *yao* (see entry no. 12, above) and *hu* (see entry no. 15, below), *ying* birds were used in hunting, especially along the northern frontiers.[135] Several early works, now lost, carry the character *ying* in their titles. Valuable information is also found in the *Youyang zazu*[136] and the relevant expertise circulated to Korea and Japan. One special account with ancient roots is a book called *Shinshu takakyo* 新修鷹経, now preserved in the library of Waseda University.[137]

Today the term *ying* stands for the *Accipitridae* family. Modern accounts often list more than twenty-five species in connection with Hainan. Some of these belong to the genus *Circus* (see entry no. 12, above), five are grouped under the genus *Accipiter*. The modern names of the latter all carry the element *ying*: *A. badius* (*he'er ying* 褐耳鷹, shikra), *A. soloensis* (*chifu ying* 赤腹鷹; Horsefield's goshawk, Chinese goshawk or sparrow hawk), *A. trivirgatus* (*fengtou ying* 鳳頭鷹, crested goshawk), *A. nisus* (*que ying* 雀鷹, Eurasian sparrow hawk), and *A. virgatus* (*songque ying* 松雀鷹, Besra sparrow hawk).[138]

The authors of the *QTZ* recognised two different kinds, as we saw: a black one and a dark brown variety (倉 read like 蒼), similar to an "owl" (see entry no. 38, below). The description of the latter is of no help; most Hainanese *Accipiter* birds have a brownish colour pattern, not dissimilar from that of a *chi*, or "standard" owl. The first item raises doubts: *Accipiter* birds residing on the island

---

134 See *Taiping yulan*, IV, j. 926, 1a–5b (pp. 4113–4115), quoting *Xijing zaji* and other materials. Another Song source: *Piya*, II, j. 16a–17a.

135 At least one late Qing chronicle of Hainan refers to the domestication of *ying* birds, but perhaps this should be read as a general statement and not related to Hainan (?). See *(Xianfeng) Wenchang xianzhi*, j. 2, p. 79. Other references in late Hainan chronicles are shorter and less explicit; see, for example, *Yazhou zhi*, j. 4, p. 83; *(Minguo) Gan'en xianzhi*, j. 4, p. 90.

136 See *Youyang zazu*, j. 20 (section called *Roujue bu*), pp. 108–111. Also see note 115, above.

137 This text, in three *juan*, seems to go back to the early ninth century. For the text, see http://archive.wul.waseda.ac.jp/kosho/wo10/wo10_00596/wo10_00596.pdf (August 2014).

138 See, for example, *Hainan dao de niao shou*, pp. 51–65 (pp. 54–58: *Accipiter* birds); Shi Haitao et al., *Hainan luqi jizhui dongwu jiansuo*, pp. 113–117.

are neither black nor completely dark. Such a colour pattern – very dark, black / brown – can best be associated with *Ictinaetus malayensis* (*lin diao* 林鵰, black eagle), a large animal also at home in these southern regions, albeit only in small numbers. Since the *QTZ* carries no special entry on the "eagle", now usually *diao* 鵰 in Chinese (formerly often *diao* 雕), of which there are several kinds on Hainan, one may not totally exclude the possiblity that, perhaps, *ying* referred to these animals as well.

(15) *Hu* / Falcons or Others

鶻: 俗名鴉鶻。

Translation: Its common name is *yahu* 鴉鶻 (literally: crow falcon).

Comment: The character 鶻 has two pronunciations: *gu* and *hu*. The version *gu* is normally linked to the combination *guzhou* 鶻鵃, often defined as the name of a dark / black bird with a short tail. This bird already appears in the *Erya*. Several commentators have equated it with the *banjiu* 斑鳩 ("turtle dove"), others say that it should be identical with the *diao* (usually "eagle", see previous entry) and / or *sun* 隼 (or 鶽, mostly "falcon").[139] The second pronunciation (*hu*) is almost exclusively associated with the *sun*.

Here the character 鶻 should be read *hu*. One intuitive explanation is that the common name *yahu* suggests an animal with a ferocious nature. The other reason lies in the fact that some local gezetteers of later periods define the *hu* as a *yao* 鷂, or praise its speed, strength and other skills; these characteristics are typical for *yao* birds (see entry no. 12, above).[140]

There can be no doubt, the name *hu* stood for birds of prey, but it appears in very different contexts.[141] Some sources define these birds as evil creatures, others stress their positive nature. Moreover, they were also used for hunting. For instance, the Northern Qi (550–577) distinguished between three kinds of *hu*. Generally, the ones with a dominating white colour were the best, the yellow ones as well as the blue / green ones came next.[142]

---

139 See, for example, *Erya zhushu*, j. 10, p. 305. Also see Read, *Chinese Materia Medica: Avian Drugs*, pp. 72–73 (no. 306) and 80–86 (nos. 311–313). – More related names in Guo Fu, *Zhongguo gudai dongwuxue shi*, p. 94 no. 2.

140 See, for example, *Yazhou zhi*, j. 4, p. 82 (name); *(Guangxu) Ding'an xian zhi*, j. 1, p. 130 (characteristics). – Also see Xian Yuqing, "Su Shi yu Hainan dongwu," pp. 119–120.

141 See, for example, *Bencao gangmu*, IV, j. 49, pp. 2674–2675 (under *chi* 鴟), citing earlier material; Read, *Chinese Materia Medica: Avian Drugs*, pp. 85–86 (no. 314).

142 See, *Youyang zazu*, j. 20 (Joujue bu), p. 109 top.

According to Tang sources, the Uighur or Huihe 回紇 people once proposed they should be called *Huihu* 回鶻 – evidently, because they admired the *hu's* extraordinary speed and power.[143] For similar reasons the *hu* became a symbol of encouragement among the military.[144] It also entered poetry, various stories and other literary texts. Zhang Jiuling 張九齡 (678–740) provides a fine example: Besides turning this bird into a hero with exceptional qualities, the author also suggests that officers and soldiers should serve their country with the determination of a *hu*.[145]

Other texts underline the noble behaviour of these birds. According to Liu Zongyuan 柳宗元 (773–819), they would hold small birds in their claws during winter nights to keep the feet warm, but in the morning they would set the prey free and disappear in the opposite direction. The conclusion is that *hu* birds were able to overcome natural desires – and above all, to show gratitude to their benefactors (the ones granting warmth). The message is clear: Individuals who are quiet, generous and docile should be preferred over those who wish to display a tough and cruel character.[146]

Hu birds were also endowed with supernatural power. The *Lingwai daida* (1178) records a *ling hu* 靈鶻 living in a tree hole, similar to a woodpecker. Once, the entrance to this cavity was blocked, but the bird caused the pluck to come out by performing a number of "ritual" steps on the ground. A man who had observed the procedure, wanted to disclose its magic; however, the bird managed to outwit him. This story reminds of a description mentioned below, in connection with the *zhuomu* 啄木 (entry no. 29). In other texts we read that a *hu* would never attack a "pregnant bird". One such reference is contained in a local gazetteer related to Wenchang.[147] In sum, the *hu* was a multi-facetted creature with very special traits.

Although the *QTZ* offers no description of the *hu*, this name should mostly refer to falcons, now usually called *sun* 隼. Modern accounts usually list three

---

143 See, for example, *Jiu Tang shu*, XVI, j. 195, p. 5210. Generally for the Uighurs in that period: Mackerras, *The Uighur Empire*; see especially pp. 97, 108, 158–159 n. 173.

144 See, for example, Han Xuehong 韓學宏, "Tang shi zhong de zhenqin shuxie" 唐詩中的珍禽書寫, pp. 30–34.

145 For the poem, see *Zhang Jiuling ji jiaozhu* 張九齡集校注, III, j. 17, pp. 910–912.

146 This is in Liu's "Hu shuo" 鶻說. See *Liu Zongyuan ji* 柳宗元全集, j. 16, pp. 137. – For the virtuous trait of the *hu*, also Read, *Chinese Materia Medica: Avian Drugs*, p. 86 (no. 313). – For an example in later Hainan chronicles: *(Xianfeng) Wenchangxian zhi*, j. 2, p. 79.

147 *Lingwai daida jiaozhu*, j. 9, pp. 372–373; Netolitzky, *Das Ling-wai tai-ta*, pp. 172–173. Shorter in Huang Zhen's 黃震 (1213–1280) *Huang shi ri chao* 黃式日鈔, j. 67, 53b, quoting *Guihai yuheng zhi* 桂海虞衡志 (1175). For this, see Hargett, *Treatises*, p. 66. – *(Xianfeng) Wenchang xianzhi*, j. 2, p. 79.

kinds for Hainan (all under the *Falconidae* family): *Falco peregrinus* (*yousun* 游隼, peregrine falcon), *F. severus* (*mengsun* 猛隼, Oriental hobby), and *F. tinnunculus* (*hongsun* 紅隼, common kestrel). The third species is quite common, the status of the other two is not clear.[148] But as was said above, in connection with the entries on *yao* and *ying* (nos. 12 and 14), traditional texts rarely make clear distinctions between falcons, hawks and other birds of prey. Therefore, in this case, one may accept the remote possibility that, besides to falcons, the term *hu* in *QTZ* referred to other birds as well.

(16) *Huangying* / Blacked-naped Orioles

黃鶯: 卽黃鸝，一名蒼庚。

Translation: [This] is the *huangli* 黃鸝, also called *canggeng* 蒼庚.

Comment: The *huangying* appears in several Guangdong and Hainan chronicles, usually under that name or under *huangli*, or just under *ying* 鶯. But these accounts rarely provide additional details.[149] Modern dictionaries list further terms most of which come from traditional sources. One work has over thirty entries.[150]

Among the early sources are the *Shi jing* (there *canggeng* 倉庚 and *huangniao* 黃鳥) and the *Erya* (*canggeng, lihuang* 鵹黃 and possibly other terms). Traditional texts also relate these birds to seasonal phenomena. But this is not discussed here.[151] As usual, one finds important material in later *leishu* compilations and *bencao* works.[152]

Several scholars have suggested one should make a distinction between the form *canggeng* (plus variants) and the terms containing the character *huang*, for yellow. The first should stand for *Oriolus chinensis* (*heichen huangli* 黑枕黃鸝, black-naped oriole), the second for *O. oriolus* (*jin huangli* 金黃鸝, or *jinying* 金

---

148 See, for example, *Hainan dao de niao shou*, pp. 65–66; Shi Haitao et al., *Hainan luqi beizhui dongwu jiansuo*, pp. 116–117. – The status of two other *sun*, *Aviceda jerdoni* and *A. leuphotes*, seems equally unclear.

149 See, for example, *(Kangxi) Qiongshan xianzhi*, j. 9, 25a (p. 542); *Yazhou zhi*, j. 4, p. 82.

150 See Sun Shu'an, *Zhongguo bowu bieming da cidian*, pp. 631–632.

151 See, for example, Legge, *The Chinese Classics*. Vol. 4: *The She King*, pp. 6, 51, 198–200, 228, 238, 264, 301–302, 418–419; Guo Fu, *Zhongguo gudai dongwuxue shi*, pp. 95 (nos. 39 and 58), 98 (nos. 69 and 73), 101 (no. 100), 111 (no. 2), 291, 339–340. – Generally, for a broader audience interested in the *Shi jing* and its birds: Yan Zhongwei 顏重威, *Shi jing li de niaolei* 詩經裡的鳥類 (Taizhong: Xiang yu wenhua, 2004).

152 See, for example, *Piya*, II, j. 8, 4b–7a; *Bencao gangmu*, IV, j. 49, p. 2658; Read, *Chinese Materia Medica: Avian Drugs*, pp. 67–68 no. 299 (*ying* 鷪).

鶯; golden oriole). Both belong to the *Oriolidae*, together with several other species, which are at home in different parts of China.[153]

While *O. chinensis* is fairly common in much of eastern China, there are conflicting views in regard to *O. oriolus*. Many works do not mention this bird in the context of China, others say it occurs in northwestern Xinjiang and only there.

The male birds of both species are yellow and black and generally look very similar, but there are variations in the size and shape of their transocular stripes. The females have some olive-green parts and are usually darker. Perhaps this explains the character *cang*.

On Hainan, one only encounters *O. chinensis*. It is mostly found in the western and southeastern regions of that island. All names in the *QTZ* should refer to this kind and not to *O. oriolus*.[154]

Today there is only one additional species of the *Oriolidae* family on Hainan: *O. traillii* (*zhu huangli* 朱黄鸝, maroon oriole), which has a very different colour pattern, as its modern name suggests. This bird is briefly mentioned under the entry no. 31, below.

(17) *Jiandao que* / Drongos

剪刀雀：类燕而大，尾長分張，善擊鴉。

Translation: It resembles a swallow in kind, but is larger; the tail is long and splits into [two] parts; it is good at striking at crows.

Commment: Identical or very similar descriptions appear in several works, for instance in Huang Zuo's *Guangdong tongzhi*. According to the *Yazhou zhi* this bird is black and its tail feathers open and close like sissors.[155] There are several other names for it, for example *wuxu* 烏鬚 and *jianmao que* 剪毛雀.[156]

---

153 Traditional paintings also show "orioles". The *Gugong niao pu*, IV, pp. 30–33, contains two illustrations and detailed descriptions. It also lists different names. One article on orioles is Leijon, "Shooting Orioles".

154 See, for example, *Hainan dao de niao shou*, pp. 192–193, which also records the names *huangying* and *huangniao*.

155 See, for example, *(Jiajing) Guangdong tongzhi*, II, j. 24, 12a (p. 630); *Yazhou zhi*, j. 4, p. 79. Also see *(Jiajing) Guangdong tongzhi chugao*, j. 31, 18a; *(Kangxi) Wenchang xianzhi* (康熙) 文昌縣志, j. 9, p. 213; *(Daoguang) Guangdong tongzhi*, III, j. 99, p. 249 top; *(Xuantong) Ding'an xianzhi*, j. 1, p. 105; *(Guangxu) Ding'an xianzhi*, j. 1, p. 131; *(Minguo) Gan'en xianzhi*, j. 4, p. 90.

156 For *wuxu*: *(Xuantong) Ding'an xianzhi*, j. 1, p. 104; *(Guangxu) Ding'an xianzhi*, j. 1, p. 130; *(Jiaqing) Chengmai xianzhi*, j. 10, p. 443; *(Guangxi) Chengmai xianzhi*, j. 1, p. 80. For

Modern accounts associate the traditional names *wuxu gong* 烏鬚公 and *jiandao yan* 剪刀雁 with *Dicrurus macrocercus*, now called *hei juanwei* 黑卷尾, or black drongo in English. As both the contemporary Chinese and English names suggest, this bird is almost completely black, and its tail does look similar to a swallow's tail (one is tempted to replace 雁 in *jiandao yan* by 燕, but the first character is correct). Moreover, some sources have characterised the behaviour of *D. macrocercus* as agressive, which could be in line with the last part of the entry in *QTZ*. A further name of this bird is *dadan niao* 大膽鳥, literally the "bird of great courage".[157] There are also various older terms that were brought into connection with the black drongo, but these are of no importance here.[158]

D. marcrocercus, already described by Swinhoe under that heading, is very common on Hainan and belongs to the *Dicruridae* family.[159] Several other members of this group live on the island, including *D. paradiseus* (*da panwei* 大盤尾, greater raquet-tailed drongo). One may be tempted to refer the description in *Yazhou zhi* to these black beauties – because of the long tail shafts –, but that could be wrong as well. The short text in *QTZ* is even more vague. It specifies no colour and, theoretically, could stand for several species, not only for *D. macrocercus*.[160]

Here we have to consider two further terms found in traditional sources: *daijian niao* 帶箭鳥 (literally: arrow-carrier-bird) and *wufeng* 烏鳳 (literally: black phoenix). According to the *Lingbiao lu yi* 嶺表錄異, quoted, for example, by Huang Zuo, the *daijian niao* looks like a "wild sparrow" (*yeque* 野雀). Its wings are yellow and green; there are two "branches" at its tail, with feathers resembling those of an arrow, hence the bird's name. The second name, *wufeng*, occurs in Song works. The associated descriptions suggest a bird similar to a magpie, but dark purple and green in colour, with a cap on its head. Again the two long tail shafts are said to be adorned with feathers. Clearly, the colours given in these accounts raise questions, but it is likely that the texts refer to *D. paradiseus* (which has a cap) and / or *D. remifer* (*xiao panwei* 小盤尾, lesser

---

jianmao que: (Kangxi) Qiongshan xianzhi, j. 9, 25a (p. 542). For a different context: *Sancai zaoyi*, j. 25, 4a (p. 412).

157 See, for example, *Hainan dao de niao shou*, p. 194. For additional names see under http://www.niaolin.net/birdbook/b672.htm (July 2014).

158 See, for example, Read, "Chinese Materia Medica: Avian Drugs", pp. 64–65 (no. 295 A); Schafer, *Golden Peaches*, p. 103 and p. 305 n. 115: *pijia* 批頰 (鵊), *beijia* 鵯鵊, *piji* 鵯鶋. Note: now and then different transcriptions; for example, in Guo Fu, *Zhongguo gudai dongwuxue shi*, p. 94 no. 4.

159 Swinhoe, "On the Ornithology of Hainan"; pp. 244–245.

160 For this family on Hainan, see *Hainan dao de niao shou*, pp. 194–199.

raquet-tailed drongo). Today, the old name *daijian niao* is still in use for *D. paradiseus*. While this bird does not belong to the fauna of modern Guangdong and Guangxi, the second species occurs in parts of southern Guangxi.[161] Perhaps both kinds enjoyed a broader distribution in earlier periods.

Finally, Huang Zuo makes a distinction between *jiandao que* and *daijian niao* (there are two separate entries in his text). The conclusion is that *jiandao que* should mostly, although not exclusively, refer to *D. marcrocercus*, while *daijian niao* normally stood for *D. paradiseus*. However, other works seem to put this in doubt. For instance, the *Sancai zaoyi*, which often deviates from "standard conventions", says the *jiandao que* comes from Qiongzhou. If this implies that we are looking at an animal endemic to Hainan, then *jiandao que* cannot refer to *D. macrocercus*. The last phrase in *QTZ* could in fact support that view: *shan ji ya* 善擊鴉 may evoke the image of a bird flying as fast as an arrow – a feature derived from the character *jian* 箭 in the name *daijian niao*. Its "target" is the *ya*, often a negative bird (see entry 13); hence the drongo should have a positive connotation, but that would of course require additional research.

(18) *Zhegu* / Partridges or Chinese Francolins

鷓鴣： 自呼鈎輈格磔，常南飛不北。雄，春護雌啼，占林曲。土人取雛養爲媒，對啼致之，設機以取。俗傳此鳥與蛇交，啼數則蛇必至，機者防之。《外紀》鷓鴣媒詩：

世路無端盡網羅，聲聲何苦喚哥哥！
哥哥若不爭雄長，叫破春風也奈何。

Translation: It calls itself "*gou zhou ge zhe*". Normally, it flies southbound, not towards the North. In spring, the male bird protects the female [with] shouts, occupying [territory in] the woods [while] singing. The natives take the newborn bird and rear it as a decoy; they cause it to shout [and when other birds] approach, they find ways to catch [them]. Tradition says, this birds copulates with snakes; when it shouts several [times], the snake will definitely come; fowlers should be

---

161 *Daijianniao*: see *Lingbiao lu yi*, j. zhong, 8a. German translation: Guignard, *Aufzeichnungen über die Wunder des Südens*, p. 53. Note: textual variations between different editions of *Lingbiao lu yi*. – *Wufeng*: see *Guihai yuheng zhi jiyi jiaozhu* 桂海虞衡志輯佚校注, pp. 81–82; Hargett, *Treatises*, p. 62 and n. 10; *Lingwai daida jiaozhu*, j. 9, pp. 369–370; Netolitzky, *Das Ling-wai tai-ta*, pp. 170–171. – For possible earlier references to *D. paradiseus*, see Schafer, *Golden Peaches*, pp. 103–104.

careful with this. The [*Qiongtai*] *waiji* [records] a poem [called] "Zhegu mei" 鷓
鴣媒:

> The way [through] life is full of snares and traps, for no reason,
> Why bother to call the elder brother!
> [If] he does not struggle for supremacy,
> [One's] shouts against the winds of spring will lead to nothing.

Comment: Normally the term *zhegu* stands for partridges. The alternative name with which the text begins bears a clear onomatopoetic dimension; many works record the *zhegu's* singing, which they transcribe in different ways, for example as "kee-kee-kee-karr" or "ka-ka, gu-gu, gui-gui-gui" (while bending the head) and "ga-ga" (while raising the head). The second pattern symbolizes the sounds uttered during the breeding season (spring and summer). Specialists say one can easily distinguish the *zhegu's* shouts from the shouts of other birds. Clearly, the *zhegu's* salient voice also contributed to the emergence of such forms as *xing bude ye gege* 行不得也哥哥, literally "it is indeed of no use, elder brother". Implicitly, "elder brother" could point to a strong character – i.e., to a fierce male *zhegu* fighting over females and territory. In essence, however, this phrase stands for hardship in life.[162]

Besides the poem cited here, Wang Zuo, author of the *Qiongtai waiji*, has left further verses on the *zhegu*, which take up this theme.[163] In the eyes of earlier poets the *zhegu* also evokes sad feelings. Examples are found in the works of Zheng Gu 鄭谷 (849–911) and Wei Yingwu 韋應物 (737–791), Li Bai 李白 and Liu Zongyuan 柳宗元 (already mentioned above).[164] The *zhegu's* association

---

162 See, for example, Zheng Zuoxin, *Zhongguo dongwu tupu: niaolei*, p. 53; *Hainan dao de niao shou*, p. 67. – Later gazetteers records the *zhegu's* voice in a very similar manner. See, for example, *(Qianlong) Qiongzhou fuzhi*, j. 1 xia, 97a (p. 103); *(Minguo) Gan'en xianzhi*, j. 4, p. 88; *Yazhou zhi*, j. 4, p. 80; *(Guangxu) Ding'an xianzhi*, j. 1, p. 131; *(Minguo) Danxian zhi*, j. 3, p. 190; *(Wanli) Danzhou zhi*, tianji, p. 36; *(Daoguang) Wanzhou zhi* (道光) 萬州志, j. 3, p. 295; *(Xianfeng) Wenchang zhi*, j. 2, p. 77; *(Jiaqing) Chengmai xianzhi*, j. 10, p. 444; *(Guangxi) Chengmai xianzhi*, j. 1, p. 81; *(Kangxi) Changhua xianzhi*, j. 3, p. 48; *(Qianlong) Lingshui xianzhi*, j. 1, p. 134. Even Swinhoe mentions this; see his "On the Ornithology of Hainan", pp. 359–360 (no. 128). – Older material: *Taiping yulan*, IV, j. 924, 8a–b (p. 4105); *Bencao gangmu*, IV, j. 48, p. 2619–2620. See further *Sancai zaoyi*, j. 9, 113a (p. 120). – The poem is also in *Jilei ji*, p. 223.

163 *Jilei ji*, pp. 29 and 223. Also http://www.hnszw.org.cn/data/news/2011/03/48916/ (July 2014).

164 *Zheng Gu shiji jianzhu* 鄭谷詩集箋注, j. 2, pp. 199–200; *Wei Yingwu shiji xinian jiaojian* 韋應物詩集系年校箋, j. 10, p. 547 (the authorship of this piece is uncertain; perhaps it was written by Li Qiao 李嶠); *Li Bai ji jiaozhu*, II, j. 8, pp. 584–585; *Liu Zongyuan quanji*, j. 43,

with the South, addressed in the *Qin jing* 禽經 (traditionally attributed to Shi Kuang 師曠, 572–532 BC, but certainly of much later origin) and other early texts, further stresses this layer.[165] Hearing its cry and seeing it heading towards the South, people will think of their distant homes in the Chinese heartlands. That in turn explains the emergence of certain other names such as *Yuezhi* 越雉 (Yue stands for the "remote" lands of modern Guandong and Vietnam), *huainan* 懷南, *nanke* 南客 (again, both indicate the same direction) and *suiyang* 隋陽 (the bird follows the sun).[166]

But why does the *zhegu* always move south? The simple answer is because it fears coldness. Some texts say it rarely comes out in the morning or evening; moreover, it covers itself with leaves during the night, avoids frosty weather, or, when flying under such conditions, carries a leaf to protect its body, because the cold dew might weaken its voice.[167]

Besides serving the needs of poets, *zhegu* birds fulfilled three other functions. First, they could be trained as pets; but when domesticated, their voices would suffer.[168] Second, *zhegu* meat was a declicacy. To some, grilled partridge tasted better than chicken. Third, various substances derived from the *zhegu* were used in medicine, for example, against poisonous plants and bad fungi.[169]

Finally, traditional texts tell us that different animals copulated with snakes. But the origin of the the relation between *zhegu* birds and snakes cited in *QTZ* is unclear.

---

p. 381. – Also *Guangdong xinyu*, j. 20, pp. 516–517, similar *Nanyue biji*, j. 8, 6a–b. – More information, for example, in Han Xuehong, *Jingdian Tang shi niaolei tujian*, pp. 62–64. Also see Guo Fu, *Zhongguo gudai dongwuxue shi*, pp. 435–436, where Liu Zongyuan and others are cited.

165 See, for example, *Qin jing*, 9b–10a (pp. 683–684; Upton's *A Study of the Avian Canon* was not accessible); *Lingbiao lu yi*, j. zhong, 9b; Guignard, *Aufzeichnungen*, pp. 57–58. – Another early source is the *Yiwu zhi*. Later scholars have tried to collect the *Yiwu zhi* fragments from different texts. One work resulting from these efforts is attributed to Yang Xiaoyuan 楊孝元 (Yang Fu 楊孚, 1st cent.); here, see *Yiwu zhi*, 3b. – Some later gazetteers repeat the information found in these and other early sources. See, for example, *(Minguo) Gan'en xianzhi*, j. 4, p. 88; *Yazhou zhi*, j. 4, p. 80; *(Jiajing) Guangdong tongzhi*, II, j. 24, 15a–15b (p. 632).

166 See *Qin jing*, 9b–10a (pp. 683–684); *Bencao gangmu*, IV, j. 48, p. 2619; *Guangdong xinyu*, II, j. 20, pp. 516–517.

167 As in previous note. Also see, for example, *Gujin zhu*, j. 2, p. 12. A very special explanation under the entry "Dan nan bu bei" 但南不北 in *Sancai zaoyi*, j. 9, 122b (p. 120)

168 See, for example, *Wanling ji*, j. 60, 8a–9a, j. 5, 5b; *Guangdong xinyu*, j. 20, p. 517.

169 See, for example, *Yiwu zhi*, 3b; *Xinxiu bencao*, pp. 395–396 (no. 485); *Bencao gangmu*, IV; j. 48, p. 2619; *(Jiajing) Guangdong tongzhi*, II, j. 24, 15a (p. 632).

In spite of these and other details, the term *zhegu* raises questions. Generally, *zhegu* birds can be found in many parts of southern China, especially in the Lingnan 嶺南 region. Their shapes were compared to those of wild chicken (*yeji* 野雞; mentioned in several entries, here), hens and *banjiu* 斑鳩 (entry no. 19), and their heads to those of *chun* 鶉 birds (entry no. 22). Yet the most distinctive feature of the *zhegu* should be its colour pattern. According to some texts *zhegu* birds have white dots on their chests, while their backs are covered with purple and red feathers.[170] But other texts, especially very late works such as the *Yazhou zhi* and the *Gan'en xianzhi*, provide different descriptions: the whole body of Hainanese *zhegu* has grey dots, there are also white spots on the plumage, the back is neither purple nor red.[171]

Today two kinds of *zhegu* live on Hainan. They belong to the *Phasianidae* and are called *Francolinus pintadeanus* (*zhegu* 鷓鴣, usually Chinese francolin) and *Arborophila ardens* (*Hainan shan zhegu* 海南山鷓鴣, Hainan hill partridge; also see entry no. 30, below). The descriptions in *Yazhou zhi* and *Gan'en xianzhi* seem to partly match these birds. The sound pattern indicated in *QTZ* suggests that *F. pintadeanus* should be the more likely candidate for *zhegu*.[172]

(19) *Jiu* / Pigeons or Doves

鳩：斑色，俗名斑鳩。又有青鳩、綠鳩 [卽向里老]、火鳩 [魚化]。《奇甸賦》：鱗登陸兮或化火鳩。

Translation: [This bird] has a brindle color and is commonly called *banjiu* 斑鳩. There are also blue / green *jiu*, green *jiu* [this is *xiang lilao*] and fire *jiu* [which a fish turns into]. The "Qidian fu" [says]: When scaly creatures climb ashore they may become fire *jiu*.

Comment: In the Tianyi ge version the passage in brackets "[卽向…]" and the part starting with "[魚化]…" both appear in white characters on black ink. Here we follow the punctuation of the modern version in short characters. However,

---

170 See, for example, *Lingbiao lu yi*, j. zhong, 9b; Guignard, *Aufzeichnungen*, pp. 57–58; *Xinxiu bencao*, pp. 395–396 (no. 485); *Guihai yuheng zhi jiyi jiaozhu*, pp. 86–87; Hargett, *Treatises*, p. 65; *Bencao gangmu*, j. 48, p. 2619; *(Xianfeng) Wenchang zhi*, j. 2, p. 77; Schafer, *Vermilion Bird*, pp. 240–241.

171 See *Yazhou zhi*, j. 4, p. 80; *(Minguo) Gan'en xianzhi*, j. 4, p. 88.

172 See *Hainan dao de niao shou*, pp. 67–69; *Hainan luqi jizhui dongwu jiansuo*, pp. 117–118. – For general works on these birds: Li Xiangtao, *Gamebirds of China*, pp. 58–59, 78–79, and Lei Fumin and Lu Taichun, *Zhongguo niaolei teyou zhong*, pp. 80–88.

the first bracket remains unexplained; perhaps one should read it in some local dialect.

Etymologically the character *jiu* consists of two parts: *jiu* 九 (nine) and *niao* 鳥 (bird). This may be linked to the concept of "change" (*yi* 易), which is described, for example, in the *Liezi* 列子. Simply put: *Yi* produces "one-ness", "one" leads to "seven", "seven" leads to "nine". "Nine" represents the final stage. Thereafter things gradually reverse to "one-ness", i.e., to the beginning of *yi*.[173] In that sense the transition from "scaly creature" / fish to *jiu*, as indicated in the verse (quoted from Qiu Jun), symbolizes only one of several steps.[174] Naturally, such ideas remain vague and lend themselves to literary allusions; this in turn may explain why the *jiu* appears in so many different contexts.[175]

In early times *jiu* was a generic term, "rather than a specific name of a bird".[176] It then entered different compounds such as *shijiu* 鳲鳩 and *jujiu* 雎鳩 (both in *Shi jing*) and came to be associated with both auspicious and inauspicious elements.[177] At the same time its multi-functional nature caused much terminological confusion.

The semantic difficulties begin with the *Shi jing*. This source does not only contain the combinations *jujiu* (conventionally defined as the *yuying* 魚鷹; see entry no. 52 here) and *shijiu* (cuckoo; also *bugu* 布谷, *dujian* 杜鵑, etc.; see entries no. 2 and 23) – it also seems to use the character *jiu* for the myna (*quyu* 鴝 鵒, *bage* 八哥, etc.; see entry no. 7). Endless discussions, involving the *Erya* and other early sources, were led on these and related terms. Qing scholars in particular complicated the issue by referring to further terms, for example *gujiu* 鶻 鳩, *mingjiu* 鳴鳩 and *banjiu* 斑鳩 (all used for pigeons), or when applying *jiu* to completely different birds such as the hoopoe (*jijiu* 鵖鳩, usually *daisheng* 戴 勝).[178]

---

173 See, for example, *Liezi jishi* 列子集釋, pp. 6–8; Graham, *The Book of Lieh-tzu*, p. 19. Also see notes to entry no. 22, below, which refers to changes / transformations as well.

174 For the whole poem, see, *Chongbian Qiongtai gao*, j. 22, 4b–14a, here 10b.

175 Early notes on the *jiu* are in Watters, "Chinese Notions about Pigeons and Doves", especially pp. 229–241.

176 See Lai, "Avian Identification", p. 350.

177 The *Taiping yulan* lists various sources which indicate both positive and negative connotations of the *jiu* bird. See *Taiping yulan*, IV, j. 921, 4b–8b (pp. 4086–4088). For a brief explanation, also see, Han Xuehong, *Jingdian Tang shi niaolei tujian*, p. 84.

178 See Lai, "Avian Identification", pp. 350–352. – Guo Fu, *Zhongguo gudai dongwuxue shi*, also lists numerous terms, especially in the context of the *Erya*. An interesting comment to the *Erya* is, for example, in *Erya yishu* 爾雅義疏, VIII, xia 5, 1a–4a. – One earlier source of confusion is the *Qin jing*, which makes no clear distinction between *shijiu*, *daisheng* and *bugu*. See *Qin jing*, 6b (p. 682). Many names are also found in *Piya*, II, j. 7, 4a–8b (various

In the *QTZ* the name *jiu* stands for the *banjiu* – i.e. doves or pigeons. One of the earliest texts on these birds is a piece by Fu Xian 傅咸 (239–294).[179] In ancient times *ban* 斑 also appears as *ban* 班. According to the *Qin jing* the second form expresses order, because this bird fed its babies from right to left in the morning, and in the reverse direction at night.[180] By contrast 斑 means "dotted"; hence *banjiu* is a *jiu* with speckles. Li Shizhen adds further details: the *banjiu* is also called *banzhui* 斑隹, *jinjiu* 錦鳩 and *bojiu* 鵓鳩. *Ban* 斑 and *jin* describe the plumage / colour, *zhui* indicates a short tail. Large and dotted *jiu* bear the name *zhujiu* 祝鳩, small ones without spots *zhui*. These birds are not good at building nests, but they forecast weather changes; the males call the females when it is clear; when it rains, they drive out the hens from the nest.[181]

Here we may turn to local gazetteers, which repeat some of the elements listed above, for example in regard to weather changes.[182] They also list additional names such as *jimu jiu* 雞母鳩, *niumu jiu* 牛拇鳩, *zhui* 雛, *qingzhui* 青雛, etc.[183] Furthermore, we learn that *banjiu* birds live in pairs and have a thin reddish neck. Generally, one can distinguish three kinds according to colour and size. The *qingjiu* 青鳩 is blue or black or green, the *lüjiu* 綠鳩 green, the *huojiu* 火鳩 red.[184] They are as large as *ge* 鴿 (see the entry no. 2), or smaller, or even "as large as a hen" (this applies to the *bojiu* 鵓鳩, also called *jimu jiu*).[185] Other

---

entries). – For studies on the cuckoo and "related" birds, see Meyr, *Der Kuckuck im alten China*; Lai, "Messenger of Spring and Morality"; Serruys, "A Note on the Names of the Hoopoe in Chinese and Mongol". Also see the long notes in Hoffmann, "Vogel und Mensch", pp. 62 et seq., and Mittag, "Becoming Acquainted", especially pp. 314, 320–321, 326 (for the cuckoo and *Shi jing* scholarship).

179 For this poem, the "Banjiu fu" 斑鳩賦, see, for example, *Han Wei liuchao baisan jia ji* 漢魏六朝百三家集, j. 46, 17a–b (p. 328). – There is also a "Jiu fu" 鳩賦 by Ruan Ji 阮籍 (210–263), discussed in many works. See Wu Yifeng, *Yong wu yu xu shi*, pp. 110–114; Guo Fu, *Zhonggu gudai dongwuxue shi*, p. 442.

180 See *Qin jing*, 16a–b (p. 687).

181 *Bencao gangmu*, IV, j. 49, p. 2651; Read, *Chinese Materia Medica: Avian Drugs*, pp. 61–62 (no. 291). Also see, for example, *Tong zhi* 通志, I, j. 76, p. 882 (top).

182 See, for example, *(Kangxi) Qiongshan xianzhi*, j. 9, 25b (p. 542); *(Daoguang) Wanzhou zhi*, j. 3, p. 294.

183 See *(Minguo) Gan'en xianzhi*, j. 4, p. 87; *(Xianfeng) Wenchang xianzhi*, j. 2, p. 77; *(Guangxu) Ding'an xianzhi*, j. 1, p. 130. – Some works also note the eating habits of *jiu* birds and provide details on their behaviour, but this is of no importance here.

184 See, for example, *(Qianlong) Qiongzhou fuzhi*, j. 1 xia, 98a (p. 103); *(Guangxu) Ding'an xianzhi*, j. 1, p. 130; *(Kangxi) Danzhou zhi ()*, j. 1, p. 35; *(Minguo) Danxian zhi*, j. 3, p. 191; *(Jiaqing) Chengmai xianzhi*, j. 10, p. 443.

185 See, for example, *(Minguo) Gang'en xianzhi*, j. 4, 87; *Yazhou zhi*, j. 4, p. 79.

sources indicate different colour patterns.[186] There is, for example, a white *jiu*, on which the *Taiping yulan* provides further details.[187] Cao Zhi, already quoted above (in entry no. 12), also notices their eye colour and red feet.[188]

Today one finds the term *jiu* in several species under the *Columbidae*. Of these eleven occur on Hainan: *Treron sieboldii* (*hongchi lüjiu* 紅翅綠鳩, white-bellied [wedge-tailed] green pigeon), *T. curvirostra* (*houzui lüjiu* 厚嘴綠鳩, thick-billed green pigeon), *T. bicincta* (*chengxiong lüjiu* 橙胸綠鳩, orange-breasted green pigeon), *Ducula aenea* (*lü huangjiu* 綠皇鳩, green imperial pigeon), *D. badia* ([*shan*]*huang jiu* [山]皇鳩, [mountain] imperial pigeon), *Columba punicea* (*zi lin'ge* 紫林鴿; pale-capped pigeon, purple wood pigeon), *Macropygia unchall* (*banwei juanjiu* 斑尾鵑鳩, bar-tailed cuckoo dove), *Oenopopelia tranquebarica* (also *Streptopelia tranquebarica*; *huo banjiu* 火斑鳩, red turtle dove), *S. chinensis* (*zhujing banjiu* 珠頸斑鳩, spotted dove), *S. orientalis* (*shan banjiu* 山斑鳩, Oriental or rufous turtle dove), and *Chalcophaps indica* (*lübei jinjiu* 綠背金鳩, second character also *chi* 翅; emerald dove). Some of them were already mentioned above, in the entry on *boge* (no. 2). Many of these birds can be related to the descriptions provided by local gazetteers.[189] But it is impossible to determine which animals should be identified with the *banjiu* in *QTZ*, although one is tempted to give a certain preference to those under the genus *Streptopelia* – on account of their current names. *S. tranquebarica*, which is reddish-brown (and also known under the names *hongjiu* 紅鳩, *hong jiazhui* 紅咖追, etc.), could be identical with the *huojiu* in *QTZ*. *Lüjiu* could point to one or several of the *Treron* members. However, it is very difficult to draw a clear dividing line between the *banjiu* and the *boge*. Colour could be the crucial factor, or perhaps a "status" distinction between domesticated and wild, or another element. This is why we prefer the general English translation "doves / pigeons" in both cases.

(20) *Baitou weng* / Chinese Bulbuls

白頭翁： 《外紀》詩并序： 瓊島山中白頭，春末夏初嗚呼： "總不好紙筆！"

總不好紙筆，生男無美質。

---

186 See, for example, *(Kangxi) Lin'gao xianzhi*, j. 2, p. 53.
187 See *Taiping yulan*, IV, j. 921, 4b–8b (pp. 4086–4088), especially 6b–7b.
188 See *Cao Zhi ji jiaozhu*, j. 2, pp. 225–226.
189 See *Hainan dao de niao shou*, pp. 113–123, and the listings in *Hainan luqi jizhui dongwu jiansuo*, pp. 131–133.

莫怪陶淵明，但飲杯中物。

生兒願愚魯，東坡厭聰明。

聰明反誤身，愚魯多公卿。

知一不知二，聽予山鳥聲。

Translation: A poem and its preface from the [*Qiongtai*] *waiji* [read]: The white-headed [bird] in the mountains of Qiong[zhou] Island calls out in late spring and early summer: "Papers and brush are disliked!"

Papers and brush are disliked,
When one produces a boy without qualities.
Do not blame Tao Yuanming,
Of only drinking from [this] cup.
Producing a boy – he should be stupid and rude,
[Yes,] Dongpo hated smartness.
Smartness may cause a life to be wasted –
The stupid and rude are mostly officials.
Knowing one thing, but not a second one,
You should listen to the songs of mountain birds!

Comment: The name *baitou weng* can stand for three things: for an aged man with grey hair, for a plant or herb called *Anemone chinensis* (*baitou weng hua* 白頭翁花, Chinese pulsatilla), and for a bird with a white patch on its black head, as in the present case.

The poem, also included in Wang Zuo's *Jilei ji*, mentions Tao Yuanming (365–427) and Su Dongpo. Above we had already encountered the latter. The former is equally famous and Su Shi admired him, often being inspired by his poetry.[190]

Tao Yuanming was a native of Chaisang 柴桑 (today Jiujiang 九江, Jiangxi Province) and is also known under the names Tao Qian 陶潛, Wuliu xiansheng 五柳先生 and Jingjie xiansheng 靖節先生. The Eastern Jin dynasty (317–420), during which he served as a minor official, was a period of unrest, upheaval and war. In his literary works Tao Yuanming repeatedly complained about these unfortunate circumstances and, more generally, the problem of corruption. The famous phase "*buwei wudoumi zheyao*" 不為五斗米折腰 calls for integrity and sincerity: "Do not bow down like a servant in exchange of five bushels of grain!"

---

190 For the poem: *Jilei ji*, p. 28. For Su Shi and Tao Yuanming: Davis, "Su Shih's 'Following the Ryhmes'". For both men in the Hainan context: Hargett, "Clearing the Apertures and Getting in Tune", pp. 158–163.

Eventually he resigned from office and retreated to the countryside where he dedicated much time to writing poetry.[191]

In spite of his solitude and disappointment with official life, Tao Yuanming did not turn away from his family. He had five sons and once complained about their immature development, saying they would not like paper and brush: A'shu 阿舒 (a nickname of his eldest son, 16 years) was extremely lazy. A'xuan 阿宣, just about to turn 15, still refused to learn writing. The two sons aged 13 did not even know the letters for "six" and "seven". The youngest one, almost 9 years, only thought of pears and chestnuts. Tao Qian concluded if this was Heaven's will, he should be drinking.[192]

Su Dongpo expressed similar sentiments in a piece called "Xi'er" 洗兒:[193]

人皆養子望聰明，我被聰明誤一生。
惟愿孩兒愚且魯，無災無難到公卿。

Everyone wishes [his] son to become smart and bright,
[But] my life is wasted through intelligence.
Now I wish [my] child to be foolish and rude,
May he become an official, without [exposure to] calamities and misery!

There can be no doubt that the author of *Qiongtai waiji* made use of the works by Tao Qian and Su Dongpo, when composing his own poem, but it is not clear what motivated him to link these verses to the *baitou weng*. Perhaps the call of this bird reminded him of the first line, hence there could be an onomatopoetic background (?). Or he simply associated the bird's name with an experienced old man with a good sense of humor.

Most local gazetteers of later periods only record the name *baitou weng* without giving further details. But there are exceptions. According to the *Guangdong tongzhi chugao* this bird is larger than a "sparrow" (*que* 雀) and has white dots on its head. A Qing chronicle of Chenghai 澄海 (eastern Guangdong), besides mentioning the similarity between both birds, says it has a longer tail; one

---

191 Tao Yuanming's biography is in *Jin shu* 晉書, VIII, j. 94, pp. 2460–2463. – A recent Chinese monograph on this man: Li Changzhi 李長之, *Tao Yuanming zhuan lun* 陶淵明傳論. A modern annotated version of his oeuvre: *Tao Yuanming ji jianzhu* 陶淵明集箋注. – Works in European languages abound; here are three examples: Hinton (tr.), *The Selected Poems of T'ao Ch'ien*; Pohl (tr.), *Der Pfirsichblütenquell*; Davis, *T'ao Yüan-ming, AD 365–427. His Works and Their Meaning*. – Also see *Jilei ji*, p. 28 n. 10.

192 For the whole poem see, *Tao Yuanming ji jianzhu*, j. 3, pp. 304–306.

193 For the poem, which is not in *Su Shi quanji*, see *Su Shi shiji*, VIII, j. 47, pp 2535–2536. For a translation and interpretation, see Egan, *Word, Image, and Deed*, p. 250. See further *Jilei ji*, p. 28 n. 11.

also calls it *baitou ke* 白頭客 (literally: the white-headed guest) because of its white "cap", and it utters sounds during the fifth watch of the night.[194]

In 1870, Swinhoe provided further information on the *baitou weng*, which he then called *Ixus hainanus*. Knowing of earlier textual references to that bird, he wonders why he only encountered "black-caps" on the island, and no "white-bonnets" (i.e., birds with a light-coloured crown), adding he had seen both kinds on the Leizhou peninsula, opposite of Hainan, at a short distance from the island (these birds he called *I. sinensis*). To solve the puzzle, he offers two tentative conclusions: (1) *I. hainanus* originated from a different location; it gradually adjusted to the insular environment by changing its appearance; thereafter it spread to the Leizhou Peninsula due to its powerful reproduction. (2) Or it was a native of Hainan and certain nearby areas, including the Leizhou region. The black-capped species emerged later. As it was better suited to the nature of the island, it began to substitute the "white-caps", which disappeared in the course of time, while their Chinese name survived.[195] – Clearly, these suppositions are not persuasive, since both white-capped and black-headed birds are found in other parts of China as well and since their reproduction is strong.

Modern scholarship equated Swinhoe's *I. sinensis* with *Pycnonotus sinensis* (*baitou bei* 白頭鵯; light-vented bulbul, Chinese bulbul), often defining the Hainanese "variety" as a subspecies: *P. s. hainanus*. Moreover, some texts still associate the old name *baitou weng* with that bird.[196] Photographs of *P. sinensis* usually show an animal with a black head, white speckles around the auriculars and a tail slightly longer than that of "ordinary" sparrows. These observations support the equation *P. sinensis* = *baitou weng*.

But there is more to say. The *Pycnonotidae* to which the Chinese bulbul belongs constitute a large family. Zoological handbooks record four other members of that group on Hainan: *Criniger pallidus* (also called *Alophoixos pallidus*; *baihou guanbei* 白喉冠鵯, white-throated bulbul), *Hypsipetes mcclellandii* (*lüchi duanjiao bei* 綠翅短腳鵯, green-winged bulbul), *H. flavala* (also *Hemixos castanonotus*; *subei duanjiao bei* 粟背短腳鵯, the chestnut-backed or chestnut bulbul), and *H. leucocephalus* (until recently also *H. madagascariensis*; *hei duanjiao bei* 黑短腳鵯, black bulbul).[197] The first three have black heads; clearly, this contradicts the idea of a *baitou weng*. The colour pattern of the fourth bird can be very different: On Hainan *H. leucocephalus* is nearly completely black; on

---

194 See *(Jiajing) Guangdong tongzhi chugao*, j. 31, 18a; *(Jiaqing) Chenghai xianzhi* (嘉慶) 澄海縣志, j. 24, 3b (p. 288). Partly similar and other details in *Gugong niao pu*, IV, pp. 36–37.

195 See, Swinhoe, "On the Ornithology of Hainan", pp. 253–255 (no. 75).

196 For example, *Hainan dao de niao shou*, pp. 183–184.

197 For example, *Hainan dao de niao shou*, pp. 183–188; Shi Haitao et al., *Hainan luqi beizhui dongwu jiansuo*, pp. 153–154.

the continent it sometimes has a white head. This probably contributed to the invention of such terms as *baitou weng* and *baitou gong* 白頭公 – however, not on Hainan, but in Guangdong and other mainland locations, where these names certainly also came in use for *P. sinensis*.[198] The conclusion could be that the nomenclature was imported to Hainan where the combination *baitou weng* was restricted to *P. sinensis*.

(21) *Baishe* / Common Blackbirds or Others

百舌：春初鳴，作百鳥聲。《月令》云：小暑至，反舌無聲。

Translation: It sings in early spring, making the sounds of one hundred birds. The "Yueling" says: "With the beginning of the *xiaoshu* 小暑 [solar period] (roughly: mid-July), it 'turns' the tongue and remains quiet."

Comment: Many ancient texts refer to the *baishe* in the same way, by quoting from the "Yueling" and mentioning this birds's extraordinary vocal talents. The *baishe* also appears in poetry, for example by Mei Yaochen 梅堯臣 (1002–1060), a famous Song scholar, who composed verses on various avian species.[199] Furthermore, *baishe* birds provide substances used in traditional medicine.

In modern writing the *baishe* normally stands for *Turdus merula* (*wudong* 烏鶇), the common blackbird (Eurasian blackbird, etc.). Zoological accounts place this bird under the *Turdinae* subfamily (which forms part of the *Muscicapidae*) or the *Turdidae* family. There are many other popular Chinese names for *T. merula*.[200] This includes *fanshe* 反舌, *niushi niao* 牛屎鳥 and *niushi bage* 牛屎八哥, literally "cowdung myna". The element *bage* in the last name can be explained on account of this bird's voice and dark colour, which makes it similar to the myna described above. Moreover, there should be a relation between the term *niushi bage* (also *niu bage*) and the form *niubei liao*, as was already mentioned under the entry *quyu* (no. 7).

---

198 The names *baimian gong* 白面公, *bailian shanque* 白臉山雀, etc. sound similar but refer to *Parus major* (*dashan que* 大山雀, great tit). See *Hainan dao de niao shou*, p. 263.

199 See *Li ji jin zhu jin yi* 禮記今註今譯, I, 275–276; Couvreur, *Mémoires*, I, pp. 359–360. Generally on birds in the "Yueling": Taylor, "'Guan, guan' Cries the Osprey", pp. 4–6. For Mei Yaochen: *Wanling ji*, j. 4, 1b–2a. There are many studies on this man, for example Chaves, *Mei Yao-ch'en and the Development of Early Sung Poetry*, and Leimbiegler, *Mei Yao-ch'en (1002–1060). Versuch einer literarischen und politischen Deutung*.

200 See, for example, http://niaolei.org.cn/search?d=baike&q=commoner&p=8 (May 2014). One also finds traditional illustrations; see, for example, *Gugong niao pu*, III, pp. 72–77 (there, the terms *bei baishe* 北百舌 and *nan baishe* 南百舌).

But one must be careful. Although the *T. merula* option is the most likely and conventional one, some authors have equated the *baishe* with *Cettia diphone* (also / earlier *Horornis cantans*, *H. diphone*, *C. cantans*; Japanese / singing bush warbler).[201] This bird, named *shuying* 樹鶯 or *Riben shuying* 日本樹鶯 in China, belongs to the *Sylviinae* (again a subfamily under the *Muscicapidae*), the *Sylviidae*, or the *Cettiidae*, depending on the classification that one wishes to follow.

Both *C. diphone* and *T. merula* exist on Hainan. The local "variants" are usually classified as *C. d. canturians* and *T. m. mandarinus*. These birds spend the winter on the island; therefore, the calendrical observations in *QTZ* are inappropriate, or rather, one should apply them to the continent.[202]

Interestingly, Swinhoe associated the term *baishe* with the "shama", calling its Hainanese "variety" *Cittacincla macrura minor*.[203] This bird was later equated with the *baiyao queju* 白腰鵲鴝, or white-rumped shama, now usually named *Copsychus malabaricus*. Its Hainanese subspecies is the *C. m. minor*. *C. malabaricus* is similar to another bird: *C. saularis*, the magpie robin, or *queju* 鵲鴝. The Hainanese subspecies of the latter bears the name *C. s. prosthopellus*. The *Copsychus* birds belong to the *Turdinae* subfamily / *Turdidae* family as well. Modern works list several popular names for *C. saularis*, but not *baishe*. Moreover, the issue of subspecies is quite complex.[204]

(22) *Anchun* / Common Quails

鵪鶉： 《列子》云：蛙變爲鶉，又云：鼠亦爲鶉。土人重山呼、㓥鷄，不取爲鬪。

Translation: The *Liezi* says, frogs turn into *chun* 鶉 birds. One also says, field mice become *chun*. The native people value *shanhu* 山呼 and *chengji* 㓥鷄 [birds] and do not use [the *chun*] for fighting.

Comment: Today the combination *anchun* stands for quails. But in old texts, the two characters appear separately and not as a binom. We shall look at the name *chun* first. This term can already be found in very early sources such as the *Shanhai jing* and *Shi jing*.[205] The graph for *chun* is also present in the compound

---

201 See, for example, Read, *Chinese Materia Medica: Avian Drugs*, p. 66 (no. 297).

202 See, for example, *Hainan dao de niao shou*, pp. 225, 244–245.

203 See Swinhoe, "On the Ornithology of Hainan", p. 344.

204 See, for example, *Hainan dao de niao shou*, pp. 215–217.

205 For example, *Shanhai jing jiaozhu*, Xishan, pp. 25, 29, 47 (associated with Di 帝), Zhongshan, p. 141; Legge, *The Chinese Classics*. Vol. 4: *The She King*, pp. 80, 171, 358; *Erya yi*, II, j. 15, pp. 161–162. – For a brief discussion of the *chun* in *Shi jing*: Fan Limei 范麗梅,

chunhuo 鶉火, which represents a stellar constellation. Besides that, the *chun* bird was brought into connection with the *liao* 鷚, the male of which goes as *jie* (*gai*) 鴰, while the female bears the name *bei* (*pi*) 庳. Furthermore, the *chun* gives birth to the *wen* 鳼 (a young quail), whereas the *ru* 鴽 (a small bird like the quail) produces the *ning* 鸋 (usually: wren). Alternatively, *wen* was also equated with as a female *shanchun* 山鶉 (partridge / birds under the genus *Perdix*), and *ning* with a female *anchun*. Others believe the *ru* should be identical with the *moumu* 鴾母 (usually defined as a crested ibis), which in turn stands for the *an*. Many of these terms, as well as several additional expressions, appear in the context of discussions surrounding the *Erya* and its later comments. Their interpretation raises innumerable questions and in several cases one even encounters different *pinyin* transcriptions. In short, the terminology is extremely complex.[206]

The process of metamorphosis mentioned in *QTZ* appears in many texts, not only in the *Liezi*. Claims that *chun* birds would originate from frogs or shrimps are common; other works say that field rats (*shu* 鼠) would turn into *ru* / *an*. One well-known source for such "transitions" is the *Da Dai Li ji* 大戴禮記.[207]

Later records provide additional details. One example is the *Taiping yulan*; it contains an entry on *yan* 鷃 birds – usually thought to be identical with *an* – and a separate one on the *chun*. The *Piya* carries a separate section on *an* and one on *chun*.[208] Huang Zuo's *Guangdong tongzhi* of the Ming dynasty has an entry on *chun*, but not on *an* or *anchun*.[209] This and other evidence suggests a continued distinction between *yan* / *an* and *chun*. Li Shizhen's *Bencao gangmu* is no exception in that regard – it carries two separate segments as well, one on *yan* 鷃 (in lieu of *an*, but which mentions this term), and one on *chun*.[210] The *chun*, we learn

---

"Yuedu 'chun zhi benben': *Shi jing* yinyong yu zhujie de duoceng quanshi" 閱讀《鶉之奔奔》：詩經引用與注解的多層詮釋. – Also see, for example, Guo Fu, *Zhongguo dudai dongwuxue shi*, pp. 44 (no. 76), 46 (no. 92), 112 (no. 26).

206 See, for example, *Erya zhushu*, j. 6, p. 176; j. 10, pp. 310–311, 313–314, 320; *Erya yishu*, VIII, xia 5, 11b–12a, 14b, 17a–b, 28a; Guo Fu, *Zhongguo gudai dongwuxue shi*, pp. 53, 96 (no. 31), 97 (nos. 39, 48, 52), 100 (no. 99).

207 See, for example, *Liezi jishi*, p. 15; Graham, *The Book of Lieh-tzu*, pp. 21–22; *Da Dai Li ji buzhu* 大戴禮記補注, j. 2, 48. Also see Guo Fu, *Zhongguo gudai dongwuxue shi*, pp. 349, 441–442. For a broader analysis of the conceptual dimensions, see Sterckx, *The Animal and the Daemon*, chapters 5 and 6, especially pp. 174, 201, 203.

208 *Taiping yulan*, IV, j. 921, 8b–9a (pp. 4088–4089), and j. 924, 9b–10b (p. 4106); *Piya*, II, j. 8, 9a–b, 12b–13b.

209 *(Jiajing) Guangdong tongzhi*, II, j. 24, 15a (p. 632).

210 For this and the following details, see *Bencao gangmu*, IV. j. 48, pp. 2622–2623; Read, *Chinese Materia Medica: Avian Drugs*, pp. 45–46 (nos. 278 and 279). One of Li Shizhen's sources is a text by Kou Zongshi 寇宗奭. See his *Bencao yan yi* 本草衍義, j. 16, p. 113.

from his book, is as large as a "young chicken, with a narrow head, no tail, dotted plumage, and a fat [body]. The feet of the males are high, those of the females are short. They fear coldness, stay in fields and wild [terrain], at night they fly in groups, during the day they hide in the grass..."

As to the differences between the *an / yan* and *chun*, the most obvious one concerns the plumage of both birds: the former has no spots on its feathers, whereas the second does. This in turn can be related to the process of meta-morphosis: one "species" comes from frogs etc., the other from rats etc. Besides such details one also encounters observations related to seasonal changes of the plumage, which has led to further names.

Many sources embed the idea of metamorphosis in a complex cosmological / Daoist setting. One example, related to the *chun*, is found in Zhou Qufei's *Ling-wai daida*; it alludes to the relativity of things by reminding readers of the famous *kunpeng* 鵾鵬 story in *Zhuangzi* 莊子.[211] Although the *QTZ* does not address such points openly, the authors and editors of local chronicles were certainly familiar with these concepts.[212]

Here we may return to the issue of names. It is not clear when the characters *yan / an* and *chun* first began to form one compound. Several sources list both terms in separate entries, as was mentioned. These entries often appear in one textual segment; moreover, some authors associated the expressions *an / yan* and *chun* with similar birds under the same "family" – in spite of minor differences in their alleged origin and appearance. In addition, related combinations such as the expression *ruchun* 鴽鶉 in *Qin jing* certainly had something to do with the emergence of the final version *anchun*.[213]

There is still more to tell. *Anchun* birds were trained to fight and their meat became a delicacy in the cuisine. Qu Dajun provides valuable remarks on the first issue, to mention just one Guangdong source. He also says, the male bird has dotted wings, black claws and long feet; the eyes of the female are yellow, the bill is red, its feet are shorter. Besides such descriptions, there are references to the methods of catching such animals.[214] Additional information is found in a

---

211 See *Lingwai daida jiaozhu*, j. 9, pp. 377–378; Netolitzky, *Das Ling-wai tai-ta*, p. 175. The relevant entry is called "chunzi" 鶉子. There is a second reference to the *chunzi* in the "chenji" 枕鷄 section (*Lingwai...*, p. 382; Netolitzky, pp. 177–178). Some scholars believe the *chenji* stands for *Coturnix coturnix* (*C. japonica*, etc.).

212 See *(Wanli) Qiongzhou fuzhi*, j. 3, 98b (p. 76); *(Qianlong) Qiongzhou fuzhi*, j. 1 xia, p. 103; *(Guangxu) Ding'an xianzhi*, j. 1, p. 131; *(Minguo) Danxian zhi*, j. 3, p. 189; *(Kangxi) Wen-chang xian zhi*, j. 9, p. 213. Some of the relevant entries in these and other local gazetteers are shorter than or slightly different from the one in *QTZ*.

213 *Qin jing*, 14a (p. 686).

214 See *Guangdong xinyu*, j. 20, pp. 526–527; *Qing yi lu* 清異錄, in *Shuo fu* 說郛, IX, j. 61, 31a.

short treatise of the Qing period. This work by Cheng Shilin 程石鄰 explains how to raise and keep the *anchun*, and how to turn them into good fighters.[215]

In view of these details one wonders why the Hainanese would not select these creatures for avian contests – unlike the *shanhu* and *chengji* (see entries no. 5 and 39). Were the local people mostly interested in *anchun* meat and eggs? – One of the earliest texts with references to the use of *chun* and *ru* in cooking is the *Li ji* ("Neize" chapter).[216] Later sources tell us that *anchun* birds were sold on the markets.[217] Huang Zuo explains one could salt and grill their meat.[218]

Modern zoology subordinates the *anchun* bird – "quail" in English, as was mentioned – to the *Phasianidae* family.[219] Normally, two kinds are mentioned in the context of Hainan: *Coturnix coturnix* / *C. c. japonica* / *C. japonica* (*anchun*; common quail, Japanese quail, Asian migratory quail) and *C. chinensis* (*lanxiong chun* 藍胸鶉; blue-chested quail, Asian blue quail). In recent writing *C. japonica* has become the preferred name option for the first bird, but some works stay with the subspecies. The past and present status of *C. chinensis* on Hainan is unclear; some sources say it may be extinct, but we cannot tell, whether it was ever widely distributed on the island. This could suggest that the *QTZ* refers to the first bird only and not so much to the second kind; however, the text provides no details that might allow us to define a particular species.[220]

(23) *Zigui* / **Cuckoos**

子規: 一名杜鵑, 聲鳴謝豹。

Translation: Another name [of this bird] is *dujuan* 杜鵑. [Its] sound [gives] Xie Pao 謝豹.

Comment: The *zigui* / *dujuan* – also called (1) *zijuan* 子鵑, (2) *zijuan* (or *zigui*) 子嶲, (3) *duyu* 杜宇, (4) *bugu* 布谷, etc. – stands for a cuckoo. Some texts link

---

215 See *Anchun pu* 鵪鶉譜. More on this work and later texts, including bibliographical details, for example, in Siebert, *Pulu*, pp. 167–168 and n. 307–309, pp. 277–278.

216 See, for example, *Li ji jin zhu jin yi*, I, 460; Couvreur, *Mémoires*, I, p. 644.

217 For example, *Erya yi*, II, j. 15, p. 160.

218 *(Jiajing) Guangdong tongzhi*, II, j. 24, 15a (p. 632).

219 For overviews, see, for example, Li Xiangtao, *Gamebirds of China*, pp. 65–69; Song Dong-liang 宋東亮 et al., "Anchun de zhonglei, fenbu, tezheng ji jiazhi" 鵪鶉的種類、分布、特征及價值.

220 See *Hainan dao de niao shou*, pp. 68–69; Shi Haitao et al., *Hainan luqi jizhui dongwu jian-suo*, pp. 117–118; http://avibase.bsc-eoc.org/species.jsp?avibaseid=C2579CCB48D8AF6F (August 2014).

the shouts of this bird, often heard through the night, to the hardship of separation; farmers have compared these shouts to the sound of a hoe, which explains yet another name: *chu'nou que* 鋤耨雀. The name *zijuan* (1) appears in the context of a story related to Wangdi 望帝, a ruler of Shu 蜀. Wangdi is also known as Du Yu (3), from which is derived the form *dujuan*. Furthermore, the *zigui / dujuan* has entered poetry. Li Shizhen and many other texts provide additional details; there are also several works, which narrate different tales and legends associated with this bird.[221]

One such story tells us of a man who came to drink in the house of a certain Xie 謝. His daughter observed him and felt quite happy. But upon hearing the shout of a *zigui*, the man felt moved, thanked his host and left the place (謝去), which made the young lady feel upset. Later, when she herself heard the *zigui*, she would imagine the roaring sound of a panther / leopard (*pao* 豹) and ask her maid to chase the beast away with a bamboo stick, saying: "*Pao*, you still dare to come here and yell?" Hence the expression "Xie Pao", which has become a further name of the *zigui*.[222]

Without doubt, the term *zigui* stands for one or several species under the *Cuculidae* family, called *Dujuan* in modern Chinese zoology. On Hainan alone there are more than ten such birds. Experts say, they can recognize the voices of these animals, but for the untrained ear that should be very difficult; hence local avian names with "onomatopetic" elements such as *kegu* 喀咕 are not listed here. The geographical distribution of the *Cuculidae* across Hainan varies from one kind to the next.[223] Since the *QTZ* and most other chronicles pertaining to the en-

---

221 Relevant studies in European languages: Meyr, *Der Kuckuck im alten China*, especially part 3.1; Lai, "Messenger of Spring and Morality", especially pp. 540–541; Hoffmann, "Vogel und Mensch", pp. 70 et seq. – Poetry: Du Fu's 杜甫 (712–770) "Dujuan xing" 杜鵑行 is one example; see sources above and Mittag, "Becoming Acquainted", p. 314. – Guangdong chronicles contain further data; see, for example, Ptak, "Zhuhai dongwu", pp. 176–177 (nos. 17 and 19). Interestingly, the *(Jiajing) Guangdong tongzhi*, II, j. 24, 16b (p. 632) gives *Hainan niao* 海南鳥 as an alternative name / classification of the *dujuan*. – There are also several names similar to *bugu*; these go back to very ancient times and link to different themes. – For Li Shizhen: *Bencao gangmu*, IV, j. 49, pp. 2665–2666; Read, *Chinese Materia Medica: Avian Drugs*, p. 63 (no. 293) and p. 73 (no. 306). – For traditional stories related to the *dujuan / zigui* etc. one may also consult "http://niaolei.org.cn/posts/3953" (June 2014).

222 This story, originally found in a work called *Chengdu jiushi* 成都舊事, is frequently quoted in *leishu* works, for example in *Ge zhi jing yuan* 格致鏡原, j. 87, 5b. Other *leishu* give the name, but provide no or a different context. One case: *Sancai zaoyi*, j. 9, 122b (p. 120); j. 24, 15a (p. 393). – For the *pao*, see Bocci, "Il leopardo nell'antica Cina".

223 See, for example, *Hainan dao de niao shou*, pp. 124–133; Shi Haitao et al., *Hainan luqi beizhui dongwu jiansuo*, p. 133–135.

tire island, and not just to one district or region, provide no additional details that might be needed for species identification, one cannot tell which bird is meant in our text. Finally, there could be a relation between the *zigui* bird and the *huoji*, described below (entry no. 30); possibly this second animal also belongs to the *Cuculidae* family.

(24) *Huamei* / Hwameis

畫眉：白眉褐質，善鳴好鬥。

Translation: It has white brows, a dark brown body, is good at singing and loves to fight.

Comment: Today the term *huamei* stands for two things: (a) for *Garrulax canorus* – in English: hwamei, Chinese hwamei or laughing thrush (also written laughingthrush) – a species still placed under the *Timaliinae* sub-family some years ago, then put directly under the *Timaliidae* family, and finally under the *Leiothrichidae* (here the *huamei* comes out as *Leucodioptron canorum*); (b) for the said sub-family (under the *Muscicapidae* or *Weng ke* 鶲科), or for the family (on the same level with the latter). Generally, traditional sources compare "ordinary" *huamei* birds with the mandarin duck, certainly because the latter, as well as several species under the *Muscicapidae* / *Leiothricidae* have a white stripe above their eyes.[224] The fighting strength of the *huamei* is also praised in many accounts. Above, attention was already drawn to a long description in *Guangdong xinyu* (see entry no. 5).[225]

Modern works pertaining to Hainan confirm that *G. canorus*, when put in cages, can be trained to fight.[226] The colours of these small creatures, which are

---

224 But some sources simply carry a description identical to the one in *QTZ* (white brows, dark brown body), without explicitly drawing a comparison. One example: *(Jiajing) Guangdong tongzhi*, II, j. 24, 12a (p. 630). For colours and brows also: *Sancai zaoyi*, j. 9, 61b (p. 94). For a comparison: *(Guangdong) tongzhi chugao*, j. 31, 17b. – European observers drew attention to the bird's "eyebrows" as well; see, for example, Hoffmann, "Vogel und Mensch", pp. 50 and 74. For a beautiful illustration of the Qing period: *Gugong niao pu*, III, pp. 30–31.

225 *Guangdong xinyu*, j. 20, pp. 517–518. There are two Qing texts on *huamei* birds: Chen Jun 陳均, *Huamei bitan* 畫眉筆談 (see: http://ctext.org/library.pl?if=en&res=82821), and a longer work by Jin Wenjin 金文錦, *Huamei jie* 畫眉解 (see: http://digital.staatsbibliothek-berlin.de/werkansicht/?PPN=PPN3343781517&PHYSID=PHYS_0006; both August 2014). Also Siebert, *Pulu*, p. 167 n. 307, pp. 264, 284. The second work in particular is very interesting; it deals with raising, feeding and training these birds.

226 See, for example, *Hainan dao de niao shou*, pp. 234–236; Lei Fumin 雷富民 and Lu Taichun 盧汰春, *Zhongguo niaolei teyou zhong* 中國鳥類特有種, p. 465.

also described as excellent singers, match the ones given in *QTZ*. But the matter is not so simple, because occasionally traditional sources confuse *huamei* and *shanhu* birds. Moreover, zoologists have found many variations within the *G. canorus* cluster. Therefore the hwamei of Hainan are often treated separately, as a subspecies called *G. c. owstoni*.[227]

To round off these remarks: On Hainan one finds several birds under the *Timaliinae* and three other sub-families which belong to the *Muscicapidae* (or the more recent taxomic categories mentioned above). It is very likely that our text refers to *G. canorus* or its Hainanese variety, but one cannot totally exclude the possibility that other birds were occasionally called by the name *huamei* as well. At the same time one may recall that, evidently, the *QTZ* distinguishes between two basic kinds: *huamei* (most likely *G. canorus*) and *shanhu* (most likely *Garrulax chinensis*; see entry no. 5).

(25) *Jiaotian'er* / Lesser Skylarks

叫天兒：似雀而大，色如之。性善鳴，愈鳴飛愈高。

Translation: It resembles the "sparrow", but is larger, the colour being the same. By nature it is good at singing; the more it sings, the higher it flies.

Comment: As was said, the term *que* may refer to sparrows or small birds more generally (see entry no. 3). But such a vague "classification", as well as the observation regarding the vocal talents of the *jiaotian'er*, help us very little in identifying this animal. Later texts often list the *jiaotian'er* without further comment or with an explanation similar to the one in *QTZ*. Moreover, the name occurs near the *huamei* and / or other birds known for their unusual sounds and voices. Tang Xianzu, already mentioned above, also follows that pattern in one of his celebrated poems.[228]

However, perhaps the *jiaotian'er* is identical with the *bantianjiao* 半天叫, mainly described in later material. These works say the *bantianjiao* is as large as a *waque* 瓦雀 (see next entry) and of a yellow or brownish colour. It lives on

---

227 See, for example, Tu and Liu Severinghaus, "Geographic Variation of the Highly Complex Hwamei (*Garrulax canorus*) Songs" (http://zoolstud.sinica.edu.tw/Journals/43.3/629.pdf; May 2014); or Lei Fumin and Lu Taichun, *Zhongguo niaolei teyou zhong*, pp. 463–465.

228 See, for example, *(Wanli) Qiongzhou fuzhi*, j. 3, 98b (p. 76); *(Jiajing) Guangdong tongzhi chugao*, j. 31, 17b. – Also see *Yiyu – A Sixteenth Century Sino-Mongol Glossary* (no author, place and date), p. 64 no. 523, on http://doktori.btk.elte.hu/lingv/apatoczky/diss.pdf (14 June 2014). – For Tang Xianzu: *Tang Xianzu shi wen ji*, I, j. 11, p. 431.

sandy spots near rivers and in fields. On a bright day it flies up and down, singing. The locals raise its offsprings.[229]

Modern dictionaries equate the term *jiaotian'er* with *jiaotianzi* 叫天子, *gaotianzi* 告天子, *gaotianniao* 告天鳥, *alandui* 阿濫堆, *tianque* 天雀, *yunque* 雲雀, etc.[230] This is based on traditional works, including various old comments to the *Erya*. The *Erya* identifies the *liu* 鷚 with a bird called *tianyue* (or *tianyao*) 天鸙. Guo Pu adds, "east of the Jiang" (Jiangdong 江東) its name is *tianliu* 天鷚. Hao Yixing 郝懿行 (1757–1824), a Qing commentator, gives the popular form *tianque* for this bird. He also says, when flying up into the sky it makes a sharp sound, as if complaining (*gaosu* 告訴), hence the name *gaotianniao*.[231] Other works, like the *Sancai zaoyi*, make a distinction between some of these names or add minor details. There is, for example, a brown *gaotianzi* and a *wuse jiaotian* 五色叫天; the latter is also called *gaotianzi* and found in Yunnan.[232]

The terms mentioned above suggest two kinds of birds: pipits / wagtails and larks / skylarks.[233] There are several pipits in China; their scientific names carry the character *liu* and they all belong to the genus *Anthus* under the *Motacillidae* (also see entry no. 45). Four of them live on Hainan. But it is not clear whether the expression *jiaotian'er* represents one or several of these birds; the *QTZ* only tells us they resemble "ordinary" sparrows (*que*).[234]

Skylarks and larks belong to the *Alaudidae*; one of these birds, *Alauda gulgula* (*xiao yunque*, 小雲雀; lesser / small / Oriental skylark), also stays on Hainan. It is brown, similar to *Passer montanus* (see entry no. 3), but slightly larger in size. Modern ornithological works such as the *Hainan dao de niao shou* provide various names for *A. gulgula*, including some of the terms listed above. Although we were unable to find the form *jiaotian'er* in the context of Hainan, the behaviour and appearance of that bird make it a good candidate for the *jiaotian'er* – a better one indeed than the pipits. One may add that Swinhoe had already noted *A. gulgula* in his research on Hainan, proposing to call it *A. sala*.[235]

---

229 See, for example, *Yazhou zhi*, j. 4, p. 83; *(Minguo) Gan'en xianzhi*, j. 4, p. 91.

230 For example, Sun Shu'an, *Zhongguo bowu bieming da cidian*, p. 75. The *Gugong niao pu*, IV, pp. 70–76, shows several *alan* birds (written 阿蘭), but their identification raises questions, partly because the pictures contradict the associated descriptions in many ways.

231 *Erya yishu*, VIII, xia 5, 4b. Also see, for example, Guo Fu, *Zhongguo gudai dongwuxue shi*, p. 94 no. 9; http://ctext.org/dictionary.pl?if=en&char=%E9%B8%99 (Chinese text project); http://niaolei.org.cn/posts/7563 (both August 2014).

232 See *Sancai zaoyi*, j. 24, 5a (p. 388), and j. 28, 19a (p. 481).

233 See, for example, Guo Fu, *Zhongguo gudai dongwuxue shi*, p. 94 no. 9.

234 See *Hainan dao de niao shou*, pp. 176–177.

235 See *Hainan dao de niao shou*, pp. 167–168, and Swinhoe, "On the Ornithology of Hainan", p. 354. – Some internet sources relate the term *jiaotianniao* 叫天鳥 to *xiao yunque*, espe-

(26) *Waque* / Sparrows

瓦雀 (no text)

Comment: The name *waque* is phonetically similar to *maque* 麻雀, usually *Passer montanus*, and should perhaps be derived from the fact that sparrows often stay under the eaves of a house with roof tiles, or on the roof itself. Besides being interchangeable with *maque*, the combination *waque* can also substitute the simple term *que* (for all this, see entries nos. 3 and 25). There are, of course, many other names for the *waque*, one being *chiyan* 斥鷃, in the first *Zhuangzi* chapter.[236]

Several later Hainan chronicles list the expression *waque*. Usually they provide no descriptions of that bird, like the *QTZ*, but now and then they offer some details. For instance, according to the *Yazhou zhi* these birds are brown; they nest near houses and twitter to celebrate happy events.[237]

Traditional illustrations are a further source. One famous painting of the Song period, which is easily accessed through the internet, bears the title "Waque qi zhi tu" 瓦雀棲枝圖; the birds shown on it can be identified as *P. montanus*.[238] This confirms the existence of two terms for one and the same species. Yet, perhaps one should be more careful in the case of our text: It is possible that *waque* and *maque* were interchangeable on two levels: on the level of individual animals, as we saw; and in a "generic" sense, i.e., both were in use for a variety of small birds similar in shape, but not necessarily with identical colours.

Here one may add a related point: The *QTZ* lists several *que*. These do not occur in a single cluster; on the contrary, they are scattered over the entire bird section. On the one hand, this seems inconsistent; on the other hand, similar accounts with comparable bird lists are not different in that regard. The *ji* 鷄 category is a further case which does not form a single block. Whether the authors and editors of Ming chronicles noticed such "problems" and what they thought of them, is a difficult question that would merit a separate treatment.

---

cially in the context of Taiwan and the Penghu Islands, near Taiwan. See, for example, http://lohas.phhcc.gov.tw/Articles.asp?mpmid=5&mpsid=3&lmenuid=2&smenuid=3&tmenuid=0 (August 2014).

236 For *waque* / *que*, see, for example, *Bencao gangmu*, IV, j. 48, pp. 2626–2631; Read, *Chinese Materia Medica: Avian Drugs*, p. 49 (no. 283).

237 *Yazhou zhi*, j. 4, p. 81.

238 See, for example, http://www.dpm.org.cn/shtml/117/@/7341.html (August 2014). This piece is now in the Palace Museum.

(27) *Bolao* / Shrikes

博勞：卽鵙，俗名馬伯勞。

Translation: [This] is the *ju* 鵙, commonly called *mabolao* 馬伯勞.

Comment: The name of this bird, often translated as "shrike", was already mentioned above, under *quyu*. The *Bencao gangmu* equates the term *bolao*, written 伯勞, with several other expressions. Among these names one also finds *ju* (also transcribed differently), *boliao* 伯鷯, etc.[239] The *bolao* is frequently compared to the myna as we saw above (see entry no. 7). The element *liao* in the last name seems to favour such comparisons.

Some older sources define the *bolao* as an unfilial bird: *buxiaoniao* 不孝鳥. They say it comes in summer and disappears in winter. The roots of its negative image remain unclear, but the fact that it begins to sing during a time of the year, when the Yin element gradually becomes active, could be one reason. The *Li ji* ("Yueling" chapter) in this context refers to the *ju* 鵙.[240] Generally, there are many stories surrounding the *ju / bolao*, which are easily found in *leishu* works such as the *Taiping yulan*.[241]

Today the term *bolao* 伯勞 stands for the *Laniidae* family. Modern works list *L. cristatus* (*huwen bolao* 虎紋伯勞, *hongwei bolao* 紅尾伯勞; red-tailed shrike, brown shrike) and *L. schach* (*zongbei bolao* 棕背伯勞; rufous-backed, long-tailed or black-headed shrike) in the context of Hainan. We shall look at *L. cristatus* first. Among its popular names are *xiao mabolao* 小馬伯勞 and *gaoju* 縞鵙. Zoological works indicate several subspecies, but there are contradictory views. Nevertheless, it seems that these birds spend the winter on Hainan, which implies that the observations cited above should refer to continental China, where the situation is reverse.

---

239 See *Bencao gangmu*, IV, j. 49, pp. 2653–2655, and Read, *Chinese Materia Medica: Avian Drugs*, p. 64 (no. 295). Another term is *gu'e*, see entry no. 34, here.

240 *Li ji jin zhu jin yi*, I, 276; Couvreur, *Mémoires*, I, p. 360. – However, there are different calendrical cycles. See, for example, Guo Fu, *Zhongguo gudai dongwuxue shi*, pp. 265, 291–292, 340, and sources mentioned there.

241 One story surrounding the *bolao* (also *bailao* 百勞) features Bo Qi 伯奇, known for his filial behaviour, but later murdered. This theme entered several works. For an example, see Reed, *A Tang Miscellany*, p. 120. Internet sources provide further stories; for a collection see: http://www.fjbirds.org/bbs/viewthread.php?action=printable&tid=4068 (June 2014). – Also see *Taiping yulan*, IV, j. 923, 5b–7a (pp. 4098–4099). – Other *leishu* have short entries. One curios case in *Sancai zaoyi*, j. 9, 64a (p. 96).

L. schach is different. Its tail and parts of the head and wings are black, the body is mostly brown with some pink. Some ornithological accounts list a Hainanese subspecies named *L. s. hainanus*, which remains on the island all around the year. Its colours are slightly different from those of the Guangdong variety. *L. s. hainanus* has been described as an aggressive creature. During the mating season in particular it becomes very noisy and imitates the sounds of other birds. Popular names for *L. schach* include *da mabolao (er)* 大馬伯勞(兒) and *Hainan ju* 海南鴠.[242]

The above makes it difficult to decide which shrike is meant in *QTZ*. The names or name elements *ju* and *mabolao* leave both options open.

(28) *Tutu*

屠突：《方輿志》：雨則先鳴。

Translation: [According to the *Qionghai*] *fangyu zhi*, [the *tutu*] sings before it rains.

Comment: The combination *tutu* is very rare and an internet search yields no satisfactory results. The *Sancai zaoyi*, a Qing work, says this: "a bird's name in Qiongzhou; small but impressive; flies up suddenly..." But there is no further description.[243]

Perhaps, then, one should look for phonetically similar names. One popular name of *Lanius cristatus*, described in the previous entry, is *tuhu bolao* 土虎伯勞.[244] There is also a shorter version: *tu bolao*. Could there be a relation between the elements *tuhu* and *tutu*?

A second candidate is the *doudou que* 兜兜雀, mentioned in *Guangdong xinyu*. This name derives from the bird's voice (*doudou*). One also calls it *jidiao que* 吉吊雀. It is similar to but larger than a *quyu* 鴝鵒 (myna, see entry no. 7), has a green / blue head and azure quills / feathers, one finds it in Dongguan 東莞 (a district in Central Guangdong) and when it comes out, there will be wind. Identical information appears, for example, in *Nanyue biji* 南越筆記. *Jidiao* seems to stand for *Garrulax perspicillatus* (*heilian saomei* 黑臉噪鶥, masked or

---

242 See, for example, *Hainan dao de niao shou*, pp. 189–192; Shi Haitao et al., *Hainan luqi beizhui dongwu jiansuo*, pp. 154–155. – Popular names are also found in internet sources, for example, http://emsbri.ac.cn/modules/client/db_detail_in.php?id=500 and http://www.dadi pedia.com/wiki/list/id/58198 (both July 2014).

243 See *Sancai zaoyi*, j. 27, 62a (p. 467). The last line may be an allusion to "Qi ao" 淇奧 in *Shi jing*. See Legge, *The Chinese Classics*. Vol. 4: *The She King*, p. 93.

244 See, for example, *Hainan dao de niao shou*, p. 189.

spectacled laughing thrush), a bird in Guangdong, but not on Hainan, and with different colours; moreover, it is smaller than a myna.[245]

Although the reference to wind (not rain) could be a vague hint to a possible relation between the *tutu* and the *doudou*, all other elements cause problems. The first option (*tuhu* = *tutu*) would imply the existence of two consecutive entries on shrikes; although both seem to represent small and very agile birds, there is nothing else to support such a view.

(29) *Zhuomu* / Woodpeckers

啄木：《本草》：褐爲雌，班爲雄，穿木食蠹。《方輿志》：觜長硬如鐵，巢枯木中。人塞其門，即以觜書符於沙泥中開之。人有効之者。

Translation: [According to] the *Bencao* the brown ones are the female birds, the males are patterned; they pierce wood and eat grubs. The [*Qionghai*] *fangyu zhi* [says]: [Their] bills are long and hard like iron, they nest inside dried trees. [When] people seal the entrance, they use the bill to draw a spell on the ground, [thus causing the entrance] to reopen. There are people who imitate this.

Comment: Identical or very similar descriptions appear in several Hainanese chronicles of the Wanli period, but these works do not necessarily mention the *Bencao* and the *Qionghai fangyu zhi*.[246] *Bencao* in our text should refer to a work predating Li Shizhen's famous *Bencao gangmu* (just as in the first entry, above); a very similar passage is also included in other material such as the fragmentary *Yiwu zhi* 異物志.[247] The second account cited in *QTZ* is now lost, as was already explained.

But there is more to say in regard to Li Shizhen's compendium. This source provides a long entry on the *zhuomu* birds with citations from earlier material. In Read's words: The *Bencao gangmu* "refers to the ability of these birds to write Chinese characters, by which they induce insects to come out. Followers of the black arts use them as spells and cure children who have poisonous boils by frightening them in these symbols."[248] The magic skills of the *zhuomu* birds are a

---

245 See *Guangdong xinyu*, j. 20, p. 521; *Nanyue biji*, j. 8, 9b. For *jidiao* see, for example, http://www.sibagu.com/china/timaliidae.html (August 2014).

246 See, for example, *(Wanli) Qiongzhou fuzhi*, j. 3, 98b–99a (p. 76); *(Wanli) Danzhou zhi*, tian-ji, p. 36.

247 Quoted, for example, in *Bencao gangmu*, IV, j. 49, p. 2659.

248 *Bencao gangmu*, IV, j. 49, pp. 2658–2660, especially p. 2659; Read, *Chinese Materia Medica: Avian Drugs*, p. 68 (no. 300).

recurrent theme and will certainly remind readers of the *ling hu* in an earlier entry, above (see no. 15).

Usually the term *zhuomu* (first character in old texts also 斸) is translated as woodpecker. Woodpeckers belong to the *Picidae* family of which several species live on Hainan.[249] The attributes given in *QTZ* (brown plus female; patterned or variegated / striped plus male) are vague and insufficient to identify a single kind. "Brown" could point to *Jynx torquilla* (*yilie* 蟻鴷, wryneck) and *Micropternus brachyurus* (also *Celeus brachyurus*; *li zhuomuniao* 栗啄木鳥, rufous wood-pecker), as indeed to several female birds; "patterned" to *Dendrocopos canica-pillus* (also *Picoides canicapillus*; *xingtou zhuomuniao* 星頭啄木鳥, grey-crown pigmy or grey-capped woodpecker) and *D. major* (also *P. major*; *daban zhuomu-niao* 大斑啄木鳥, greater pied or great spotted woodpecker), to mention just some possibilities. The last bird in particular has a distinctive white and black pattern and a red patch on its nape. The female one is mostly black and lacks the red part. *Picus canus* (*heizhen lü zhuomuniao* 黑枕綠啄木鳥 or *huitou lü zhuo-muniao* 灰頭綠啄木鳥; black-naped green or grey-faced woodpecker), *P. flavinucha* (*da huangguan [lü] zhuomuniao* 大黃冠[綠]啄木鳥; greater or large yellow-naped woodpecker) and its smaller "version", namely *P. chlorolophus* (*huangguan [lü] zhuomuniao* 黃冠[綠]啄木鳥, small or lesser yellow-naped woodpecker), are mostly green or olive, with some grey. *Blythicipus pyrrhotis* (*huangzui zao zhuomuniao* 黃嘴噪啄木鳥 or *huangzui li zhuomuniao* 黃嘴栗啄木鳥; [red-eared] bay woodpecker) is perhaps the least likely candidate due to the fact that the male bird is mostly reddish-brown. Finally there is *Megalaima oorti* (*heimei ni zhuomuniao* 黑眉擬啄木鳥, *shan ni zhuomuniao* 山擬啄木鳥, *heimei nilie* 黑眉擬鴷, etc.; black-browed or Malayan barbet, etc.; also see entry no. 35, below). Today zoologists place this bird under the *Megalaimidae* (and no longer under the *Capitonidae*), sometimes calling it *M. faber* (*Zhonghua ni zhuo-muniao* 中華擬啄木鳥, Chinese barbet). It is mostly green and easily confused with certain other birds.

Qing sources pertaining to Hainan repeat some of the details found in *QTZ*, but also report new elements, for example, that there would be a small and green variety, as large as a *que* 雀.[250] Perhaps this refers to one of the *Picus* birds or *M. oorti*.

---

249 For the following, see *Hainan dao de niao shou*, pp. 156–164; Shi Haitao et al., *Hainan luqi beizhui dongwu jiansuo*, pp. 142–143 (the description of colours varies from earlier work!). – Note: Some modern texts list subspecies; these were not named here.

250 See, for example, *(Daoguang) Qiongzhou fuzhi*, j. 5, 42b.

(30) *Huoji* / "Fire-chicken"

火鷄：紅色，觜翅俱黑。春夜先鷄而鳴，俗亦謂之催耕。

Translation: [This bird] is red in colour, with the bill and wings both black. In spring nights it is the first fowl to cry. Commonly one also calls it *cuigeng* 催耕.

Comment: In many texts the term *huoji*, literally "fire-chicken", stands either for a turkey, or, as some scholars believe, for imported cassowaries. Turkeys come from the Americas; in the early sixteenth century they were not yet known in China. Cassowaries are mostly found on the large island of New Guinea. Early Ming texts related to the voyages of Zheng He 鄭和, when speaking of *huoji* in the context of Southeast Asian ports, probably refer to this bird. Furthermore, the element "fire" in its name certainly reminded some authors of the ostrich, which had a reputation of swallowing flames and burning objects.[251] Needless to point out – in the context of early Hainan, the term *huoji* should refer to a different animal and not to cassowaries, ostriches, or turkeys.

The term *cuigeng* usually designates the cuckoo (see entries 19 and 23). But it is not clear when and why the *huoji* came to be equated with this bird. Other Hainanese chronicles often provide similar information without giving further details. Following the *QTZ*, they mention two distinct features: first, the *huoji's* shout is heard earlier than the cock's crow; second, the red-and-black colour pattern. To this one may add two further observations: the *huoji* is mostly present in contexts related to southern China and it belongs to the *ji* "category", which is not too different from the modern *Phasianidae* family. However, among the *Phasianidae* of Hainan there is no single species with identical colours.[252]

Only one candidate comes close to the brief description in *QTZ*: the famous *Arborophila ardens* (*Hainan shan zhegu* 海南山鷓鴣, "Hainan hill partridge"; also see entry no. 18, above). This bird has a black head and bill and a few dark feathers on its wings; its breast is mostly chestnut-coloured with some red parts. Moreover, its habitat is restricted to Hainan.[253] Could this imply that we are

---

251 For details and selected sources, see, for example, Ptak, "Chinese Bird Imports from Maritime Southeast Asia", pp. 202–207, and "The Avifauna of Macau: A Note on the *Aomen jilüe*", section VI. Also see Zhou Yunzhong 周運中, "Hetuo yu Alabo ma" 鶴鴕與阿拉伯馬, p. 308.

252 See, for example, the descriptions in *Hainan dao de niao shou*, pp. 66–74.

253 This bird has been the object of thorough research. For general portrays, see, for example, Li Xiangtao, *Gamebirds of China*, pp. 78–79; Lei Fumin and Lu Taichun, *Zhongguo niaolei teyou zhong*, pp. 80–88.

looking at a Hainanese *huoji*, which was different from a southern Chinese continental one? Did the continental *huoji* stand for a species of the genus *Tragopan*, to mention just one possibility?

Above we saw that the *huoji* also appears in the *shanhu* chapter (entry no. 5). There it features as an aggressive animal eating other birds. By contrast, the Hainan hill partridge, which mostly lives on plants, should be a "peaceful" animal. But there is no other *ji* on the island that fullfills all the different criteria reported in *QTZ*. Perhaps, then, something is wrong with our text. Alternatively, the editors confused certain details or knew of strange legends from which they concluded that a red bird, called "fire-chicken", should have a fierce nature and kill its rivals.

Further confusion stems from the fact that Swinhoe lists the term *huoji* in his chapter on *Centropus rufipennis*, the crow-pheasant (or common coucal), now usually called *Centropus sinensis*, or *hechi yajuan* 褐翅鴉鵑 in Chinese, a bird under the *Cuculidae* family.[254] This bird is found across much of southern China, on Hainan and Taiwan, and some zoologists have defined the Hainan variety as a subspecies: *C. s. intermedius*. That also applies to *C. toulou* (or *C. bengalensis*; *xiao yajuan* 小鴉鵑; lesser crow pheasant or lesser coucal), the Hainan "version" of which bears the name *C. t. bengalensis* (if a subspecies is accepted). The colours of both species / subspecies are the reverse of the pattern indicated in *QTZ*: their bodies are mostly black and their wings are brown or chestnut-coloured.[255] Although both species are quite noisy, they are nowhere described as being agressive (as the *huoji* in the *shanhu* section). Clearly, more research will be needed to solve the *huoji* puzzle.

(31) *Hongtou ying* / "Red Orioles"

紅頭鶯 (no text)

Comment: Many of the birds described above also appear in the Wanli edition of *Qiongzhou fuzhi*. The *hongtou ying* and the following names are no exceptions in that regard, but in most cases the *Qiongzhou fuzhi* provides no explanations.[256] Other Hainan chronicles do not even list the term *hongtou ying*. Evidently this name was rare and one only finds it in one or two texts such as *Qiongshan xian-*

---

254 Swinhoe, "On the Ornithology of Hainan", pp. 234–235.
255 See, for example, *Hainan dao de niao shou*, pp. 131–133; Shi Haitao et al., *Hainan luqi beizhui dongwu jiansuo*, p. 135.
256 *(Wanli) Qiongzhou fuzhi*, j. 3, mainly 98b–99a (p. 76).

zhi of the Kangxi period and the *Wanzhou zhi* of the Daoguang era, albeit again without further comment.[257]

Today the name *hongtou (shanwei) ying* 紅頭(扇尾)鶯 is sometimes used for *Cisticola ruficeps*, the red-pale (-pate) cisticola, a bird under the *Cisticolidae* family. This bird is found in parts of Africa, but not in China. Therefore, we should look at the character *ying* instead. *Ying*, without specification, normally stands for the "oriole"; by analogy the combination *hongtou ying* could point to Swinhoe's *Psaropholus ardens*, or "red oriole".[258] This in turn links to *Oriolus trailli*, or *zhu huangli* 朱黃鸝, the maroon oriole (also *zhuli* 朱鸝, or vermilion oriole), which belongs to the *Oriolidae* family (already mentioned in entry no. 16, above). *O. traillii* is a red / maroon-coloured bird with black head and wings, but there are some colour variations within the species. The subspecies on Taiwan, *O. t. ardens*, has very luminous feathers, which does not apply to its South Asian counterparts. The Hainan variety appears as *O. t. nigellicauda* (sometimes *lise huangli* 栗色黃鸝) in the literature, but there is much confusion in regard to names.[259]

Besides these problems there emerges another very fundamental question: Local varieties of *O. trailli* have a black and not a red head, much in contrast to what the name *hongtou ying* suggests. That also applies to the male versions of *Pericrocotus flammeus* (*chihong shanjiaoniao* 赤紅山椒鳥, scarlet minivet) and *P. solaris* (*huihou shanjiaoniao* 灰喉山椒鳥, grey-throated minivet), both found on Hainan. Whether *hongtou ying* should refer to one or several of these birds, therefore, is hard to tell.

(32) *Limu que*

黎母雀： 出昌化。

Translation: It comes from Changhua 昌化 [district].

Comment: Several works associate this bird with Changhua on the western side of Hainan, but without giving further details.[260] The *Qiongzhou fuzhi* of the Qian-long reign links it to the Li Mountains 黎山 in Central Hainan, saying it would have a red crest like a flower. The *Lingao zhi* 臨高志 is given as a source for

---

257 *(Kangxi) Qiongshan xianzhi*, j. 9, 25b (p. 542); *(Daoguang) Wanzhou zhi*, j. 3, p. 295.

258 Swinhoe, "On the Ornithology of Hainan", pp. 342–343.

259 See, for example, *Hainan dao de niao shou*, p. 193. One electronic source for names is on: http://www.sibagu.com/taiwan/oriolidae.html (July 2014).

260 *(Jiajing) Guangdong tongzhi chugao*, j. 31, 18b.

these explanations.[261] Huang Zuo also quotes earlier material, listing the *limu que* in a sequence of several *que*, but provides no description of this bird.[262] More details only emerge from later works: The bird's body has a flowery pattern, like a *limu tong* 黎母桶 (seat of / or a pair of local / Limu trousers?), hence its name. The *Sancai zaoyi* says it would be a Qiongzhou bird and alludes to a local legend, without telling us how this bird looks.[263]

Evidently, the form *limu que* is indirectly related to the Limu shan 黎母山. These mountains appear in many works and are surrounded by exotic stories. Fan Chengda 范成大 gives one example: "Tradition has it that there are people up there who take comfort and pleasure in longevity and who have never had contact with the outside world."[264] The *QTZ* also refers to this terrain, but the *limu que* bird does not occur in such contexts.[265]

(33) *Suoyi he* / **Chinese Pond Herons (?)**

簑衣鶴：俗名牛奴。

Translation: Its popular name is *niunu* 牛奴.

Comment: An identical explanation is found, for example, in the chronicles of Ding'an, which assign this animal to the category "water birds". The *Wanzhou zhi* of the Kangxi period only lists the name *niunu*, without explanation.[266]

He 鶴 normally designates various kinds of "cranes". Cranes, as symbols of immortality, appear in hundreds of stories and poems, on paintings, ceramics and other objects of art, they fill long chapters in encyclopaedic works, and there is even a separate text on the these elegant animals. The origins of this mini-account, called *Xiang he jing* 相鶴經, remain obscure; according to tradition it is a very old piece, with quasi-mythological roots, that went through the hands of a

---

261 *(Qianlong) Qiongzhou fuzhi*, j. 1 xia, 96b, (p. 102). Also *(Daoguang) Qiongzhou fuzhi*, j. 5, 43b (p. 140), and other late works. – Further: *(Kangxi) Lin'gao xianzhi*, j. 2, p. 53.

262 *(Jiajing) Guangdong tongzhi*, II, j. 24, 15a (p. 632).

263 *Yazhou zhi*, j. 4, p. 81; *(Minguo) Gan'en xianzhi*, j. 4, p. 89; *Sancai zaoyi*, j. 24, 30a (p. 401). – There are very few internet entries on the *limu que* and these offer no explanation. One example is on http://www.haijiangzx.com/2012/haijiangshenghuo_1112/36399_2.html (July 2014).

264 *Guihai yuheng zhi jiyi jiaozhu*, pp. 219–220; Hargett, *Treatises*, p. 214. Also see *Lingwai daida jiaozhu*, j. 1, pp. 22–24; Netolitzky, *Das Ling-wai tai-ta*, p. 10.

265 See, for example, *QTZ*, j. 5, 19b–20b.

266 *(Xuantong) Ding'an xianzhi*, j. 1, p. 107; *(Guangxu) Ding'an xianzhi*, j. 1, p. 133; *(Kangxi) Wanzhou zhi* (康熙) 萬州志, j. shang, p. 144.

certain Wang Ziqiao 王子喬 (6th century B.C.?); later it was redone by Wang Anshi 王安石 (1077).[267]

The term *he* appears in several later Hainan chronicles. In one or two cases one finds the combination *huihe* 灰鶴 in lieu of *he*.[268] Some works also list more than one kind of "crane", for example, *he* and "*suoyi he*, commonly called *tian niunu* 田牛奴*". The *Yazhou zhi* of the late Qing period suggests three different Hainanese "species": a black and white variety with a red patch on its head, the *huihe* and the *suoyi he*.[269] Guangdong chronicles usually refer to *he* birds as well, but often provide different details. Huang Zuo complicates the issue by including the *haihe* 海鶴 in his entry on *he*; the term *haihe* appears, for example, in *Hai yu* (1534), but has little or nothing to do with "cranes".[270]

Today *huihe* usually designates the common crane, i.e., *Grus grus*. This bird belongs to the *Gruidae* family and has been reported in various parts of southern China, including Hainan, where it spends the winter months. According to modern accounts, it mostly stays in the northwestern parts of the island.[271] Since *G. grus* is the only species of its family on Hainan, the question arises, how one should interpret the terms in *QTZ* and other traditional sources related to that island.

---

267 The text is available in several collections, for example in *Wuchao xiaoshuo daguan*, vol. 9. Also see, for example, Kaiser, "Unsterblich problematisch: *Grus japonensis*", especially, p. 9; Siebert, *Pulu*, pp. 244, 295. For cranes in poetical literature, see, for example, Kroll, "Seven Rhapsodies", p. 6; Spring, "The Celebrated Cranes of Po Chü-i". For these and related themes: Guo Fu, *Zhongguo gudai dongwuxue shi*, especially pp. 173, 254–255, 436–438.

268 In one or two cases there are suprising explanations. For example, in *(Daoguang) Wanzhou zhi*, j. 3, p. 295: the *huihe* has a grey body and is red in colour! Certainly not a bird under the *Gruidae* family, see below.

269 *(Xianfeng) Wenchang zhi*, j. 2, p. 76 (*he* and *suoyihe*); *Yazhou zhi*, j. 4, p. 79. The description in *Yazhou zhi* is similar to the one in *(Daoguang) Guangdong tongzhi*, III, j. 99, p. 247 top. In each case the first part follows Fan Chengda's account. See *Guihai yuheng zhi jiyi jiaozhu*, pp. 85–86; Hargett, *Treatises*, pp. 64–65.

270 *Hai yu*, j. 2, 4a; *(Jiajing) Guangdong tongzhi*, II, j. 24, 14b (p. 631). Also see, for example, Ptak, "Chinese Bird Imports from Maritime Southeast Asia", pp. 227–228.

271 Comprehensive works on cranes in China include, for example: Li Xiaoming 李曉明 and Ma Yiqing 馬逸清, *Dandinghe yanjiu* 丹頂鶴研究; Wang Qishan 王岐山, *Zhongguo de he, yangji he bao* 中國的鶴,秧雞和鴇. – For *G. grus* and Hainan, see *Hainan dao de niao shou*, pp. 76–77; Shi Haitao et al., *Hainan luqi beizhui dongwu jiansuo*, p. 120. – Swinhoe, "On the Ornithology of Hainan", p. 362, mentions the "common crane", then called *Grus cinerea*, adding these birds, which are "very abundant" on Hainan, "largely feed on sweet potatoes" and are "prized as food by the natives". Modern accounts confirm the excellent taste of crane meat; this applies to *Grus grus*, but also, for example, to *canglu* 蒼鷺 (*Ardea cinerea* or grey heron) and other species.

The name *suoyi he* is similar to *suoyu he* 簑羽鶴. Today this second combination stands for *Anthropoides virgo*, the demoiselle crane, which is slightly smaller than *G. grus* and has an overwhelmingly grey plumage, like the latter, but does not show a red patch on its head. *A. virgo* is a migratory bird that travels over long distances from Northeast Asia to various destinations; however, zoological works do not record its presence on Hainan. This makes it impossible to equate the term *suoyi he* found in traditional Hainan chronicles with *A. virgo*.

Here one may recall a brief explanation given by Swinhoe: "The waterproof cloak of the Chinese is made of bamboo leaves, and has the appearance of the neck of the Squacco-Heron in winter plumage."[272] The term "waterproof cloak" stands for the two characters *suoyi / suoyu*. Although Swinhoe relates this combination to one single species, namely the squacco heron, it was probably a general term used for the "mantle" of various animals. Indeed, one regularly encounters this expression in modern zoological studies, where it is applied to the slender feathers hanging down on the back of certain birds. Such birds are found, for example, among the *Ardeidae*.[273]

The next term to consider here is *(tian) niunu*. This is one of many names for *Ardeola bacchus*, the Chinese pond heron, now called *chilu* 池鷺 in China. It also belongs to the *Ardeidae* family, several members of which are found on Hainan.[274] The head, neck and breast of *A. bacchus* are all chestnut-coloured. Swinhoe mentions this bird in his chapter on *Ardeolo prasinosceles*, or the squacco-heron, but retains doubts in regard to its identity.[275] Below we shall encounter another bird with the character *lu* 鷺 in its name, a further member of the *Ardeidae* group (entry no. 43, also see entry no. 44).

The suggestions offered above – *suoyi he / suoyu he* = a general term; *(tian) niunu* = *A. bacchus* – make it difficult to find an appropriate solution for the entry in *QTZ* and related references in other chronicles. If one treats *suoyi he* as a general name, the equation *suoyi he* = *niunu* should be acceptable (because the latter would be one kind under the former); this would imply the possibility of using the first term for other birds as well. Alternatively *suoyi he / niunu* may stand for one distinct species. In that case *A. bacchus* is a suitable candidate.

However, one wonders why there is no clear reference to cranes in the entire *QTZ* list. Is it possible that the author thought of "ordinary cranes", i.e., *G. grus*, when mentioning *suoyi he / niunu*? Was he unaware of the different colours of all

---

272 See Swinhoe, "On the Ornithology of Hainan", p. 365.

273 See, for example, Zhongguo yesheng... Qian Wenyan, *Zhongguo niaolei tujian*, p. 18; Shi Haitao et al., *Hainan luqi beizhui dongwu jiansuo*, p. 108.

274 See, for example, *Hainan dao de niao shou*, especially pp. 39–40: *A. bacchus / (tian) niunu*. An older scientific name is: *Buphus bacchus*.

275 See Swinhoe, "On the Ornithology of Hainan", p. 365.

these birds? Were the two terms imported and carelessly applied to the local fauna? At present there are no clear answers to these questions.

(34) *Gu'e* / Koels

姑惡 (no text)

Comment: According to the *Guangdong tongzhi chugao* this bird comes from Qiongzhou. Some texts consider the name *gu'e* as a synonym for *bolao* (see entry no. 27 here). Li Shizhen records that a women who is badly treated will turn into a *gu'e* bird. Related stories are found in other sources. They specify an unpleasant sister-in-law or mother-in-law.[276]

The *Qiongshan xianzhi* of the Xianfeng period says the *gu'e* is as big as a *que* 鵲, or magpie, completely black in colour, with red eyes and a voice that evokes sadness. This partly reminds of Li Shizhen's explanations in whose work one also finds the form *kuniao* 苦鳥 as an alternative name of this bird. The *Sancai zaoyi* lists some negative aspects, as expected, but classifies the *gu'e* as a "water bird", which may be wrong. Elsewhere it gives the name *kugu* 苦姑, derived from its voice.[277]

The *gu'e* appears in poetry as well. There are verses by Lu You 陸游 (or Lu Fangweng 陸放翁, 1125–1210), Su Shi and others, even by Wang Zuo. Furthermore, some of the descriptive elements associated with this animal in local chronicles – the black colour, the red eyes, etc. – entered the English account by Swinhoe.[278] They are found in his long chapter on *Eudynamis malayana*, i.e., the "koel", now usually called *E. scolopaceus* (*malayanus*), or Asian koel. This bird, named *zaojuan* 噪鵑 in modern Chinese zoology, occurs on Hainan and belongs to the *Cuculidae* family. As other members of that family it does not build its own nest, but uses the nests of crows and magpies, which is already stated in

---

276　*(Jiajing) Guangdong tongzhi chugao*, j. 31, 18b. – Synonym: Read, *Chinese Materia Medica: Avian Drugs*, p. 64 (no. 295), but no explanation. Also see *Bencao gangmu*, IV, j. 49, p. 2654–2655. – For a similar story, based on the *Bencao*, see *(Xuantong) Ding'an xianzhi*, j. 1, p. 105.

277　*(Xianfeng) Qiongshan xian fuzhi* (咸豐) 瓊山縣志, j. 3 xia, 43a–b; *Sancai zaoyi*, j. 9, 111b (p. 119) and j. 24, 23b (p. 397).

278　Lu Fangweng's poem, called "Xiaye zhou zhong wen shuiniao sheng shen ai ruo yue gu'e gan er zuo shi" 夏夜舟中聞水鳥聲甚哀若曰姑惡感而作詩, is in *Jiannan shigao* 劍南詩稾, j. 14, 34a (and also partly quoted in the *Qiongshan xianzhi* of the Xianfeng period, cited above). For Su Shi's verses, see *Su Shi quanji*, I, j. 20, p. 245. For Wang Zuo: *Jilei ji*, p. 28. – Also see Swinhoe, "On the Ornithology of Hainan", pp. 232–234, and the article on http://blog.sina.com.cn/s/blog_65f7afed01011e45.html (June 2014).

early sources. Its loud shouts have been "transcribed" as *sao a sao a* 嫂阿嫂阿, i.e., "sister-in-law, sister-in-law". The male bird is indeed quite black, with some lighter parts.[279]

As was said, there are several members of the *Cuculidae* on the island (also see entry nos. 23 and 30); therefore one cannot totally exclude the possibility that the name *gu'e* was occasionally used for more than one species; however, in most cases this form should refer to the koel.

(35) *Datie que* / Chinese Barbets

打鐵雀 (no text)

Comment: The name *datie que* rarely occurs in traditional bird lists and local chronicles provide very little that might be of help in identifying it. Nevertheless, Swinhoe, who translated the Chinese term as "iron smith", because "its voice sounds like hammering the metal", calls it *Megalaema faber*. This is the Chinese barbet. He also draws attention to its similarity with *M. nuchalis*, a bird associated with Taiwan and now usually called *wuse niao* 五色鳥 or *Taiwan ni zhuomuniao* 臺灣擬啄木鳥 in Chinese and Taiwan barbet in English.[280]

In early zoological works *Megalaema faber* also appears as *Psilopogon faber* and *Megalaima faber*. Generally, the taxonomy is quite confusing in this case. Later one finds the versions *P. oorti*, *M. oorti* (Malayan barbet, black-browed barbet) and other combinations. Some texts on Hainan prefer the classification *M. oorti faber*. There are several modern Chinese names for this bird of which *shan ni zhuomuniao* 山擬啄木鳥 is perhaps still the most common one.[281] Recently, however, the taxonomy has changed and *M. oorti* now appears as *Zhonghua ni zhuomuniao* 中華擬啄木鳥 (see the entry no. 29, above).

Swinhoe provides several details on this bird, which is mostly green / olive in colour, as was said, but also shows other colours near or around its head and neck. One wonders why in *QTZ* it does not appear after or before the entry on the "woodpecker".

(36) *Zhuji* / Chinese Bamboo Partridge (?)

竹鷄：自呼泥滑滑。俗傳白蟻聞之化爲水。

---

279 See, for example, *Hainan dao de niao shou*, p. 130.
280 See Swinhoe, "On the Ornithology of Hainan", pp. 96–97.
281 See, for example, *Hainan dao de niao shou*, pp. 156–157.

Translation: It calls itself "*nigugu*" 泥滑滑. Tradition says: white ants, when hearing them, will turn into water.

Comment: Several Hainan and Guangdong chronicles mention the name *zhuji*. Huang Zuo, for example, lists this bird along with various other *ji*.[282] Later sources offer many more details: The shape of the *zhuji* looks like that of a *zhegu* 鷓鴣 (for this bird see entries no. 18 and 30); it is brown in colour, with a reddish pattern, and lives in bamboo forests; the sounds it utters resemble the sequence *nigugu* (or *nihuahua*; two possible transcriptions for the last two characters); it immediately attacks its rivals. The part with the ants is found as well.[283]

Essentially, these elements also appear in Li Shizhen's work. Taken together, they should refer to *Bambusicola thoracica* (now called *huixiong zhuji* 灰胸竹鷄, the Chinese bamboo partridge), as suggested by Read (albeit with some hesitation). *B. thoracica* belongs to the *Phasianidae* family and is widely distributed in much of southern China. The name *nigugu* (or *nihuahua*), current in some regions, imitates the sounds which this bird utters.[284]

There is evidence that Chinese bamboo partridges – also called (*Hua'nan*) *zhu zhegu* (華南)竹鷓鴣, *shan junzi* 山菌子, *shanzhe* 山鷓, etc. – were trained to fight. To increase their pugnacity, people would give them small pieces of meat. This is described, for example, in *Mengxi bitan* 夢溪筆談 (dated 1086–1093), where one encounters the term *shanzhe*.[285] Below, in the entry on *shanji* 山鷄 (no. 39), we shall return to the issue of bird fights (especially "fighting chicken"; *douji* 鬥鷄).

The reference to ants in *QTZ* becomes clearer through the *Bencao gangmu*. According to that text, the *zhuji* would eat white ants (termites). Hence the saying: "The cry of a *zhuji* at home turns white ants into mud." Occasionally the *zhuji* would pick poisonous things as well, which would make these birds unsuitable for consumption. There are several stories of persons who suffered from food-

---

282 *(Jiajing) Guangdong tongzhi*, II, j. 24, 12a (p. 630).

283 See, for example, *(Qianlong) Qiongzhou fuzhi*, j. 1 xia, 95a (p. 102); *(Daoguang) Qiongzhou fuzhi*, j. 5, 41b (p. 139); *(Minguo) Danxian zhi*, j. 3, p. 188; *Guangdong xinyu*, j. 20, p. 523; *(Daoguang) Guangdong tongzhi*, III, j. 99, p. 248 top. – For a poem, by Mei Yaochen: *Wanling ji*, j. 4, 14a. A "comment" in verses, by Wang Zuo: *Jilei ji*, p. 28.

284 *Bencao gangmu*, IV, j. 48, pp. 2620–2621; Read, "*Chinese Materia Medica: Avian Drugs*", pp. 43–44 (no. 275); *Sancai zaoyi*, j. 9, 113b (p. 120), and j. 26, 11a (p. 430); Guo Fu, *Zhongguo gudai dongwuxue shi*, pp. 171, 272, 435. – Generally for this bird: Li Xiangtao, *Gamebirds of China*, pp. 82–83; Lei Fumin and Lu Taichun, *Zhongguo niaolei teyou zhong*, pp. 91–96.

285 See *Mengxi bitan jiaozheng* 夢溪筆談校證, I, j. 13, p. 462; Herrmann, *Pinsel-Unterhaltungen am Traumbach*, p. 90.

poisoning because they took too much *zhuji* meat. In such cases ginger is recommended as an antidote.[286]

Although the above provides a clear picture at first, there is a fundamental problem: Modern zoological accounts do not list *B. thoracica* among the species of Hainan. Two explanations are possible: (1) this bird disappeared from the island during the last one or two centuries, while it was still there in Ming and early Qing times; (2) the name *zhuji* was imported from the mainland and carelessly applied to a different bird similar in shape. – The fact that some Hainanese chronicles of the last dynasty list certain features which one can easily associate with *B. thoracica* (for example, its colours) could be an argument for the first option; but that would require further discussion.

(37) *Laoguan* / Storks

老鸛 (no text)

Comment: Interestingly, one of the earliest Guangdong chronicles says the *laoguan* comes from Qiongzhou.[287] Is this to suggest that we are looking at a species then not available on the mainland? – Other "southern" texts list its name, but provide no explanation.[288] In all, *laoguan* birds are rarely listed in traditional Hainan chronicles and works dealing with Guangdong in toto.[289]

The term *guan* 鸛 by itself, normally, is a general term for storks which belong to the *Ciconiidae* family. Modern accounts associate two members of that family with Hainan: *Ibis leucocephalus* or *Mycteria leucocephala* (*caiguan* 彩鸛 or *baitou huanguan* 白頭䴉鸛; painted stork) and *Leptoptilos javanicus* (*tuguan* 禿鸛, lesser adjutant). But references to these birds often remain vague and it seems that both animals can no longer be seen on the island today.[290] Perhaps, this was different in the sixteenth century; however, it is also possible that storks were already rare at that time and therefore not frequently listed in the sources.

Another possibility is that *laoguan* refers to a different species. There are several options: for example, some kind of egret or crane, or even *Gorsachius magnificus* (*Hainan huban jian* 海南虎斑鳽 or simply *Hainan jian* 海南鳽; Chi-

---

286 For example, Guo Fu, *Zhongguo gudai dongwuxue shi*, pp. 171, 272.

287 See *(Jiajing) Guangdong tongzhi chubao*, j. 31, 18b.

288 For example *(Wanli) Qiongzhou fuzhi*, j. 3, 99a (p. 76), and *(Kangxi) Qiongshan xianzhi*, j. 9, 25b (p. 542).

289 There is a special late Ming work on storks which we did not consult: Jiang Dejing 蔣德璟, *Guan jing* 鸛經. For a reference, see Siebert, *Pulu*, pp. 244, 282.

290 See, for example, *Hainan dao de niao shou*, p. 47. For detailed data on the adjutant see, for example, http://birdbase.ies.hro.or.jp/rdb/rdb_en/leptjava.pdf.

nese night heron, white-eared night heron), although the latter is easily distinguishable from a "typical" stork. Most likely, then, the authors applied the unspecific term *laoguan* to a bird which, they thought, was somehow similar to a continental stork.

(38) *Chixiao* / Owls

鴟鴞: 夜飛鳴取食。聲呼連三者不祥。俗名夜猫。

Translation: It flies at night, howling and searching for food. Three shouts in a row [mean] bad luck. Its common name is "night cat" (*yemao* 夜猫).

Comment: *Chixiao* is a general term for owls. These birds appear in Chinese mythology and there are ancient objects of art showing them in various formats. They also have entered different texts under different names. Jia Yi's 賈誼 (usually 200–168 B.C.) "Funiao fu" 鵬鳥賦 is one of the most famous works. Generally, owls are often associated with bad omina and unfortunate events. Lexicographic, encyclopaedic and other sources contain many stories of that kind. One source citing earlier material is the *Bencao gangmu*.[291]

Modern ornithological works on Hainan record several species that belong to the *Tyotinidae* and *Strigidae* families. Moreover, today the second family is called *chixiao* in Chinese.[292] But we cannot tell which species the *QTZ* describes, because there are no details concerning colour, size and general shape, head and ears tufts of the bird(s) in question. That also applies to other Hainan chronicles; they list the name, giving alternative forms as well, but add few characteristics. One exception is the Qianlong version of the *Qiongzhou fuzhi*.[293] Among other things it says these birds are as big as the *ying* 鷹 (see entry no. 14 here), yellow

---

291 See *Bencao gangmu*, IV, j. 49, pp. 2675–2677 (Jia Yi: p. 2677); Read, "*Chinese Materia Medica: Avian Drugs*", pp. 86–89 (nos. 315–316). – Eduard Erkes was one of the first European scholars to explore the cultural dimensions of China's owls. See his "Der ikonographische Charakter einiger Chou-Bronzen. II: Die Eule – und Nachtrag...". Also see Waterbury, *Bird-deities in China*, pp. 87–89; Hoffmann, "Vogel und Mensch", p. 62; Sterckx, *The Animal and the Daemon*, various references; Mittag, "Becoming Acquainted", p. 321. Recent Chinese works include Sun Xinzhou 孫新周, "Chixiao chongbai yu Huaxia lishi wenming" 鴟鴞崇拜與華夏歷史文明. – For Jia Yi, see, for example, Hightower, "Chia I's Owl Fu"; Knechtges, *Wen xuan*, III, pp. 41–49; Wu Yifeng, *Yong wu yu xu shi*, pp. 45 et seq. – For owls in poetry, also see Tan Mei Ah, "Beyond the Horizon of an Avian Fable", especially pp. 230–231.

292 See, for example, *Hainan dao de niao shou*, pp. 133–138.

293 See *(Qianlong) Qiongzhou fuzhi*, j. 1 xia, 96b (p. 102). Almost the same in *(Daoguang) Qiongzhou fuzhi*, j. 5, 44a (p. 140).

and black, with stripes (斑); they have eyes like cats and ear feathers. Males and females howl in turns, their voice is like a shout, later it sounds like laughter. Some of these features also appear in earlier sources.

The *ying*-like size, which certainly points to a large bird, and the reference to ear tufts make it possible to narrow down the options considerably, because many owls have no "ears". Both these criteria seem to be fulfilled by *Ketupa zeylonensis* (*he yuxiao* 褐魚鴞, brown fish owl) and perhaps also by *Asio flammeus* (*duaner xiao* 短耳鴞, short-eared owl), although the status of the latter on Hainan is unclear. The colours of *Otus spilocephalus* (*huangzui jiaoxiao* 黃嘴角鴞, spotted scops owl) would fit as well, but this bird is much smaller, just like *O. bakkamoena* (also *O. lettia*; *ling jiaoxiao* 領角鴞, collared scops owl). Other species, larger in size, are recorded on the mainland, but not on Hainan. That also includes *Bubo bubo* (*diaoxiao* 鵰鴞, [forest] eagle owl).

The popular name *yemao* in the *QTZ*, occasionally extended to *yemaozi* 夜貓 子 (*mao* sometimes 貓), is one of several current terms used for all kinds of owls, but especially for *B. bubo*. It is as variable as the familiar version *maotouying* 貓 頭鷹, again a generic term. On Hainan these terms seem to apply to more than one bird within the *Strigidae* family. For a further set of names, possibly related to that family, readers may consult entry no. 48, below.

(39) *Shanji* / Red Jungle Fowl

山鷄：與家鷄無異，但耳白，先鳴而後拍翼，聲較短。

Translation: [This bird] is not different from the "house chicken" (*jiaji* 家鷄), but its "ears" are white; it first shouts, then moves the wings, its sounds being quite short.

Comment: Some Hainan chronicles repeat the description found in *QTZ* almost word by word, or in part.[294] Other texts simply list the name *shanji* or mention different kinds of *ji* 鷄, saying they would be similar to the *shanji*, or even belong to that "class"; in such cases the term *shanji* represents a larger category – i.e., "mountain fowl".[295] Several sources give a very different background: the *shanji* loves its feathers; when seeing its mirror image, it begins to dance; in former

---

294 See, for example, *(Kangxi) Wenchang xianzhi*, j. 9, p. 213; *(Guangxu) Ding'an xianzhi*, j. 1, p. 131; *(Xuantong Ding'an xianzhi*, j. 1, p. 104; *(Jiaqing) Chengmai xianzhi*, j. 10, p. 444; *(Guangxu) Chengmai xianzhi*, j. 1, p. 81.

295 See, for example, *(Jiajing) Guangdong tongzhi*, II, j. 24, 10b (p. 629). This source partly relies on entries quoted in *Taiping yulan*, IV, j. 918, especially 9b–10 (p. 4074). – Also see entry no. 10, above.

time such birds were presented as gifts to the court. These elements come from the *Yi yuan* 異苑, a source quoted, for example, in *Taiping yulan*.[296] Other texts say the *shanji's* tail feathers are long.[297]

The *Bencao gangmu* lists the term *shanji* in connection with different kinds of pheasants: *dizhi* 鸐雉 and *bizhi* 鷩雉. A related example is found in *Guangdong xinyu*; it defines *jinji* 錦鷄 birds, usually equated with *Chrysolophus pictus* (*hongfu jinji* 紅腹錦鷄, the golden pheasant), as *shanji*.[298] Some works, including local chronicles, compare the *shanji* to the *baixian* (*Lophura nycthemera*, silver pheasant; see entry no. 11).[299] There are also examples for the equation of *shanji* birds with one type or two types of *zhi* (see entry no. 10), namely with the ones that have a *jinhua* or *jinqian* pattern; this partly overlaps with the *bizhi* entry in *Bencao gangmu*.[300]

These details are confusing indeed: the term *shanji* is associated with pheasants, but we also read that *shanji* were similar to or even identical with *jiaji*, i.e., domesticated chicken (as reported in our text). To get around the terminological problems, one should assume the co-existence of two different concepts with identical labels or, alternatively, open up the *shanji* "class" in a way that allows us to assign pheasants and other fowl, besides "ordinary" chicken, to one and the same category; this second option would then be close to the current *Phasianidae* concept.

Be this as it may, the similarity between *shanji* and *jiaji* is already alluded to in Shen Ying's 沈瑩 *Linhai yiwu zhi* 臨海異物志 (3rd century?), a text cited in *Taiping yulan*. Other early works, quoted in the same source, provide further details; however, the relevant elements differ from the ones given in *QTZ*. For instance, instead of mentioning "white ears", one text says *shanji* birds were black, and we also learn that they could be trained to fight.[301]

---

296 See, for example, *(Qianlong) Qiongzhou fuzhi*, j. 1 xia, 95b (p. 102); *(Daoguang) Qiongzhou fuzhi*, j. 5, 41b (p. 139); *(Qianlong) Lingshui xianzhi*, j. 1, p. 133; *Yazhou zhi*, j. 4, p. 81; *(Min'guo) Gan'en xianzhi*, j. 4, p. 89. – For the *Yi yuan* citation: *Taiping yulan*, IV, j. 918, 9b (p. 4074). – Also see Swinhoe, "On the Ornithology of Hainan", p. 359.

297 See, for example, *(Xianfeng) Wenchang xianzhi*, j. 2, p. 78.

298 *Bencao gangmu*, IV, j. 48, pp. 2616, 2617; Read, "Chinese Materia Medica: Avian Drugs", pp. 40–41 (nos. 270 and 271); *Guangdong xinyu*, j. 20, p. 522; another reference in *Sancai zaoyi*, j. 8, 65a (p. 55). – For a modern description of *C. pictus*, see Lei Fumin and Lu Taichun, *Zhongguo niaolei teyou zhong*, pp. 306–328.

299 See, for example, *(Kangxi) Lingshui xianzhi* (康熙 陵水縣志), j. 1, p. 133; *(Qianlong) Qiongzhou fuzhi*, j. 1 xia, 96a (p. 102).

300 See *(Guangxi) Ding'an xianzhi*, j. 1, p. 131; *(Xuantong) Ding'an xianzhi*, j. 1, p. 105.

301 For the *Linhai yiwu zhi* and other early sources, including the reference to the *shanji's* black colour: *Taiping yulan*, IV, j. 918, 9b–10a (p. 4074). These works are frequently cited; see,

The last point requires a brief comment because the fragments available in *Taiping yulan* suggest that *shanji* and *jiaji* were crossbred in order to obtain good "fighting fowl", i.e., *douji* 鬪鷄. Several later works contain long descriptions of such birds, appreciating their qualities and explaining how one should keep them. One key source is an anonymous text called *Ji pu* 鷄譜 (usually dated 1787), which carries many details on the *douji*. An important entry predating the *QTZ* is found, for example, in *Lingwai daida*.[302]

Occasionally *douji* birds also appear in the context of Hainan. The *QTZ* mentions them in its section on domesticated animals, under the heading *ji* (chicken), where they are equated with the *chengji* 㞯鷄. But there is no description that might help us to solve the many questions surrounding all these birds.[303]

Besides mentioning fighting fowl, the *ji* entry in *QTZ* lists several other birds: *wugu* 烏骨 (or *zhusi ji* 竹絲雞; in English often "silky fowl"; a white "house chicken"), *aijiao* 矮脚 (a short-footed chicken, now mostly raised in Guizhou), and *xuelitan* 雪裏炭 (again a white chicken, usually with black feet and head).[304] Furthermore, there are the *fanmao* 番毛 and the *chaoji* 潮鷄. The last two in particular already appear in early sources pertaining to southern China. The *chaoji*, literally "tidal fowl" (also *sichao ji* 伺潮鷄, "waiting-for-the-tide-fowl"), utters sounds to indicate the change from low to high tide; some texts say its shout is long and clear, similar to the sound of a horn. Possibly this bird is identical with the *changming ji* 長鳴鷄, which is often associated with the area of modern Vietnam. The *fanmao* (first character also 翻, 飜, etc.), "frizzled fowl" in Read's

---

for example, *(Daoguang) Guangdong tongzhi*, III, j. 99, p. 248. – For a modern note see, for example, Guo Fu, *Zhongguo gudai dongwuxue shi*, p. 272.

302 There are many modern histories of the *douji*. The early works by Wu Dachun 武大椿 are particularly well-known. See Wu's short "Douji shikao" 鬪鷄史考 and *Zhongguo douji he zawen* 中國鬪鷄和雜文 (Huhehot: Yuanfang chubanshe, 2003). *Douji* also appear in traditional verses. See, for example, Cutter, *Cao Zhi*, pp. 109–111, and *The Brush and the Spur*, which contains many references to these animals and various translations, especially from poetry. – For a modern Chinese edition of the *Ji pu*, see *Ji pu jiaoshi* 鷄譜校釋. For bibliographical information: Siebert, *Pulu*, p. 169 n. 313, p. 286. – *Lingwai daida jiaozhu*, j. 9, pp. 378–380; Netolitzky, *Das Ling-wai tai-ta*, pp. 175–177. Both refer to earlier sources as well.

303 *QTZ*, j. 9, 1b.

304 There are many modern works on the history of (domesticated) fowl in China. Early studies are Zhang Zhongge 張仲葛, "Woguo jiaqin (ji, ya, e) de qiyuan yu xunhua de lishi" 我國家禽（鷄、鴨、鵝）的起源與馴化的歷史 and Xie Chengxia, "Zhongguo jizhong de lishi yanjiu" 中國鷄種的歷史研究. The old but long entry in Read, "Chinese Materia Medica: Avian Drugs", pp. 29–38 (no. 268), is also very useful; it mentions several names listed here. Some of these names / birds are easily found in the context of traditional medicine.

translation, has "wings and tail feathers that grow in reverse direction and are curved and crooked...".[305]

The *chaoji* is also called *shiji* 石鷄. Today the term *shiji* stands for *Alectoris chukar*, the Chukar partridge, a resident of northern China, but this is of no relevance. Here we should return to the main theme of our entry because some works assign the *shiji* to the *shanji* category, quite in contrast to the classification used in *QTZ*. Huang Zuo's *Guangdong tongzhi* contains one such example; at the same time, however, this work seems to make a difference between *shiji* and *chaoji* (the latter described as a small short-necked animal associated with Guangzhou). One may add that an early reference to the "tidal fowl" / *shiji* appears in a frequently-cited *fu* by Sun Chao (Chuo) 孫綽 (314–371).[306]

As was said, the above suggests that the combination *shanji* can be both a generic term and an expression used for various individual "species", mostly found in southern China. The chapter on *Gallus ferrugineus* in Swinhoe's work seems to confirm that view. Moreover, Swinhoe describes this wild bird as having a small comb, an ochrous-brown bill, brownish-grey legs tingled with purple, "cream-white" skin "under the ear", etc.[307] The last attribute could have something to do with the "white ears" in our text.

Pictorial and other evidence from modern sources allows us to conclude that Swinhoe's description should refer to *Gallus gallus* (now *yuanji* 原鷄, usually red jungle fowl). In all likelihood this is also the animal meant in *QTZ*. There is ample evidence for the presence of *G. gallus*, sometimes called *yeji* 野鷄 (compare entry no. 10), on the island of Hainan, and indeed, many sources confirm that domesticated chicken are not too different from their "wild" (yet smaller) counterparts in the forests.[308]

---

305 See, for example, *Lingwai daida jiaozhu*, j. 9, pp. 381–383 (*changming ji, chaoji, fanmao ji*); Netolitzky, *Das Ling-wai tai-ta*, pp. 177–178; *Guihai yuheng zhi jiyi jiaozhu*, p. 84 (*fanmao ji, changming ji*); Hargett, *Treatises*, p. 64 (quotation); Read, "Chinese Materia Medica: Avian Drugs", pp. 29 and 31 (no. 268) – Three later sources: *Guangdong xinyu*, j. 20, pp. 522–523; *Nanyue biji*, j. 8, 4b–5a; *(Daoguang) Guangdong tongzhi*, III, j. 99, p. 248 top. – One of the earliest sources for the "tidal fowl" is the *Yiwu ji* 異物記 (*Yiwu zhi*) quoted in *Taiping yulan*, IV, j. 918, 8b (p. 4073). – For its long sound, see, for example, *(Kangxi) Lingshui xianzhi*, j. 1, p. 133.

306 See *(Jiajing) Guangdong tongzhi*, II, j. 24, 10b (p. 629). – The *fu* appears, for example, in *Taiping yulan*, IV, j. 918, 10b (p. 4074) and is also mentioned by Huang Zuo.

307 Swinhoe, "On the Ornithology of Hainan", pp. 357–358.

308 See, for example, *Hainan dao de niao shou*, pp. 71–72 (which lists the names *shanji* and *yeji*).

(40) *Shuiya* / "Water Ducks"

水鴨: 卽鳧鷖。

Translation: [These] are the *fu yi* 鳧鷖.

Comment: The combination *fu yi* (first character also 鳬) forms the title of a song in *Shi jing* 詩經. The first stanza of that song opens with the verse "The wild ducks and widgeons are on the Jing River" (鳧鷖在涇). Subsequent stanzas also begin with this bird combination. While, generally, *fu* is mostly equated with wild ducks, the second character remains unclear and interpretations vary substantially. One frequent explanation is "sea gull" (also see entry no. 49, below), but many specialists have suggested other readings.[309] In all likelihood later authors had no clear idea what they were referring to when citing the *Shi jing* and using the characters *fu* and *yi*. All they would be able to tell is that these were some kind of water birds.

Later Hainan chronicles often repeat the "equation" presented in *QTZ*, usually without giving further comments.[310] However, there are exceptions: One example is found in the Wenchang chronicle of the Xianfeng period; this work explains the term *fu* with *shuiya* 水鴨, adding there would also be a *haiya* 海鴨 (literally: sea duck) of slightly larger size, and a *guanya* 冠鴨 (crested duck).[311] Earlier one finds references to these and other animals in the *Guangdong xinyu*, which has four different entries on *ya* 鴨 birds.[312] Occasionally later sources also declare that *shuiya* birds would be similar to the *jiaya* 家鴨, the "house duck", but explanations are rare. Many more references are of course listed in traditional *leishu*.

Today there are several species on Hainan, which belong to the family *Anatidae*. One species, *Anas crecca*, called *lüchi ya* 綠翅鴨 in modern Chinese zoology (the common or Eurasian teal), is also known as *xiao fu* 小鳧, *xiao shuiya* 小水鴨, etc. This is as close as we can get to the statement in *QTZ*. *A. crecca*, one

---

309 See, for example, James Legge, *The Chinese Classics*. Vol. 4: *The She King*, pp. 479–481 (the translation cited here) and comments on p. 479; Guo Fu, *Zhongguo gudai dongwuxhe shi*, p. 54, no. 158, p. 113 no. 36, p. 115 no. 97; Read, "Chinese Materia Medica: Avian Drugs", p. 23 (no. 263).

310 *(Wanli) Qiongzhou fuzhi*, j. 3, 99a (p. 76); *(Qianlong) Qiongzhou fuzhi*, j. 1 xia, 97b (p. 103). The same in *(Jiajing) Guangdong tongzhi chugao*, j. 31, 18b.

311 See *(Xianfeng) Wenchang xianzhi*, j. 2, p. 78.

312 *Guangdong xinyu*, j. 20, pp. 524–525, 529.

may add, spends the winter on Hainan, especially in the northern and eastern coastal zones of the island.[313]

The above mostly concerns wild birds. But the *QTZ* also carries an entry on "ducks" (simply called *ya*) in its section on domesticated animals. This entry says one type of *ya* was called *haiya*; it was living in the wet fields and had first come to Hainan during the Chenghua period (1465–1487); later the *haiya* were gradually crossbred with domestic ducks (*jiaya*). This process led to many colour variations and, generally, to an increase in bird size.[314]

Evidently, the *QTZ* distinguishes between "wild" *shuiya* and domesticated *haiya* / *jiaya*. Some later sources seem to deviate from that pattern, as we saw. Nevertheless, to maintain the distinction between these two, we chose the translation "water ducks" for *shuiya*. Furthermore, if the characters *fu* and *yi* represent two different categories – say, *A. crecca* (plus similar birds?) and "sea gulls" (or another group of birds) – then we are looking at a larger class with two subordinated components, each of which can possibly be divided into further entities.

(41) *Bailian ji* / White-breasted Waterhens

白臉鷄: 一名鴣鷄。

Translation: It is also called *guaji* 鴣鷄.

Comment: Several Guangdong and Hainan chronicles list this bird, sometimes giving the same alternative name.[315] The *Sancai zaoyi* says it comes from Qiongzhou, Qu Dajun equates it with the *chunhun niao* 春魂鳥.[316] Modern name lists relate the latter to the cuckoo (*zigui*, etc.; see, for example, entries no. 23 and 30 here). The Daoguang version of *Qiongzhou fuzhi* places its entry on the *baimian ji* 白面鷄 (*mian* in lieu of *lian*) behind the one on the *zigui*.[317] Other works follow the same arrangement, occasionally providing short descriptions of the *baimian ji*: it lives near the water, is black, has a white face and shouts at

---

313 See, for example, *Hainan dao de niao shou*, pp. 48–51; Shi Haitao et al., *Hainan luqi beizhui dongwu jiansuo*, pp. 111–112.

314 *QTZ*, j. 9, 1b. – For domesticated ducks in Chinese history, see, for example, Zhang Zhongge, "Woguo jiaqin (ji, ya, e) de qiyuan yu xunhua de lishi".

315 See, for example, *(Jiajing) Guangdong tongzhi chugao*, j. 31, 18a; *(Wanli) Qiongzhou fuzhi*, j. 3, 99a (p. 76); *(Qionglong) Qiongzhou fuzhi*, j. 5, 25a; *(Kangxi) Qiongshan xianzhi*, j. 9, 25b (p. 542).

316 See *Sancai zaoyi*, j. 9, 71a (p. 99); *Guangdong xinyu*, j. 20, p. 523.

317 See *(Daoguang) Qiongzhou fuzhi*, j. 5, 45b (p. 141). Also, for example, *(Minguo) Danxian zhi*, j. 3, p. 192.

night.[318] But there are some variations as well: The Ding'an chronicles mention the *baimian ji* near the *zigui*, saying the former would nest in fields and their shouts would sound like *tu zu huo* 徒祖活 or *guo guo* 郭郭.[319] A Wenchang gazetteer gives *yuji* 鴒鷄 (*yu* – an error?) as an alternative form for *baimian ji* and moves the entry on the *zigui* to a different place.[320]

The term *baimian ji*, evidently a variant form of the older (?) version *bailian ji*, is one of several popular names for *Amaurornis phoenicurus* (*baixiong ku'e niao* 白胸苦惡鳥, white-breasted waterhen) under the *Rallidae* family. The habitat of this bird ranges from continental southern China to Taiwan and Hainan. Its other names include the combinations *baixiong yangji* 白胸秧鷄, *baifu yangji* 白腹秧鷄 and *bailian yangji* 白臉秧鷄, as well as the short form *ku'e niao* 苦惡鳥. Its voice comes close to the transcription *guoguo* mentioned in the Ding'an chronicles. The name *guaji*, in *QTZ*, seems to be phonetically derived from this sound, just as the version *ku'e*.[321]

The observations given in some Qing works are quite correct, indeed: During the mating season *A. phoenicurus* is noisy at night, while it usually remains calm in winter. It has a white breast and face; its back is black and it lives in rice fields and near wet areas more generally. Finally, the "proximity" of the *baimian ji* to the cuckoo could stem from the fact that the name *ku'e* (*niao*) is phonetically similar to the terms *gu'e* 姑惡 and *kuniao* 苦鳥, which represent the koel, a bird of the *Cuculidae* family (see entry no. 34). Transcriptions of the sounds it utters also remind of the cuckoo.

(42) *Feicui* / Kingfishers

翡翠：大小數種，大者毛充貢，小者呼水翠。《外紀》：宿食各占磯塘，他翠往參必見鬥。土人以媒網置占處，奮迅來擊，觸網而顛。聞諸網者云：翠性最惜其羽，罩網猶僵伏不動，恐傷羽。

Translation: There are several large and small varieties. The feathers of the larger ones [can be used as] tribute gifts, the smaller ones are called *shuicui* 水翠. The [*Qiongtai*] *waiji* [reads]: "To rest and eat, they stay near piers and ponds. [When] other *cui* 翠 show up, they will fight. The natives install a decoy and a net at [that] place; being prepared for [the birds] to rush in, they will release [the net]

---

318 See *Yazhou zhi*, j. 4, p. 82; *(Minguo) Gan'en xianzhi*, j. 4, p. 90.

319 See *(Xuantong) Ding'an xianzhi*, j. 1, p. 106; *(Guangxu) Ding'an xianzhi*, j. 1, p. 133.

320 See *(Xianfeng) Wenchang xianzhi*, j. 2, pp. 77 and 78.

321 Some names listed in *Hainan dao de niaoshou*, pp. 79–80.

and cause it to fall [over them]. Those who look after the nets are [sometimes] heard to say [this]: The *cui's* nature is to love its feathers most; caught in a net, it becomes motionless for fear it might injure them."

Comment: *Feicui* birds and feathers frequently appear in old texts, including poetry. They were submitted as tribute to the imperial court since very early times and also under the Ming dynasty. Sources related to the voyages of Zheng He contain detailed information on this. The feathers mostly came from what is now northern Vietnam, but we also know of other production sites in continental Southeast Asia and of tributes sent from maritime countries, and even from within China.[322] Local chronicles refer to the domestic trade. *Feicui* feathers were mainly used as adornments; the meat of these birds was used in medicine.

There are also various references to bird hunters. Song sources in particular, which mention *feicui* feathers in the context of Hainan and other areas, provide valuable details.[323] Among the works of the Yuan period, one may consult Zhou Daguan's 周達觀 *Zhenla fengtu ji* 眞臘風土記 (1297); it tells us that Khmer hunters were able to catch three to five birds on a good day; presumably the feathers were then sent to China. Another source says they were sold on the markets in Guangzhou.[324] The *Haicha yulu* 海槎餘錄 (1540), written two decades after the *QTZ*, links the *feicui* to Hainan's tribal areas and, once again, explains how they were obtained.[325]

---

322 For sources and other details: Ptak, *Exotische Vögel*, especially pp. 85–96; "Chinese Bird Imports from Maritime Southeast Asia", especially pp. 218–219. Some of the texts cited in the notes below are also quoted in these articles. – Earlier notes in English include Schafer, *Vermilion Bird*, pp. 238–239. – For the kingfisher in poetry, see, for example, Kroll, "The Image of the Halcyon Kingfisher".

323 For *feicui* in Song works, see *Guihai yuheng zhi jiyi jiaozhu*, p. 85 (also p. 221, p. 222 n. 1); Hargett, *Treatises*, p. 64; *Lingwai daida jiaozhu*, j. 2, p. 71; j. 9, p. 371; Netolitzky, *Das Ling-wai tai-ta*, pp. 34, 171–172; *Zhufan zhi zhubu*, pp. 433–434; Hirth / Rockhill, *Chau Ju-Kua*, pp. 183, 235–236 (minimal textual variation in Han Zhenhua's reading). – For Song imports more generally: Wheatley, "Geographical Notes on Some Commodities Involved in Sung Maritime Trade", here especially p. 99; Lin Tianwei (Lin Tien-wai) 林天蔚, *Songdai xiangyao maoyi shigao* 宋代香藥貿易史稿, table on pp. 174–208. Bielenstein, *Diplomacy and Trade*, also provides references to earlier imports.

324 *Zhenla fengtu ji jiaozhu* 眞臘風土記校注, pp. 81, 84, 93; Pelliot, *Mémoire sur les coutumes du Cambodge de Tcheou Ta-Kouan*, pp. 26, 28. – Guangzhou sales: *Dade Nanhai zhi canben* 大德南海志殘本 (preface of 1304), p. 36. Also see Yu Changsen 喻常森, *Yuandai haiwai maoyi* 元代海外貿易, pp. 133–134, quoting *Siming xuzhi* 四明續志, j. 5.

325 *Haicha yulu*, 12a. Generally on this work: Ehmke, *Das Hai-cha yu-lu als eine Beschreibung der Insel Hainan in der Ming-Zeit*.

As to the bird's special concern for its own feathers – this image already appears in a comment to the *Qin jing*.[326] Similar passages are found in more recent texts.

The *QTZ* is correct in stating that there are different *feicui* birds. Swinhoe is the first European to confirm this view with respect to Hainan.[327] China's "kingfishers" are usually grouped under three families (but there are deviations from this pattern): (1) *Alcedinidae* (*Cuiniao ke* 翠鳥科), (2) *Halcyonidae* (*Feicui ke* 翡翠科) and (3) *Cerylidae* (*Yugou ke* 魚狗科). Seven species which belong to these families live on the island: (a) *Alcedo atthis* (*putong cuiniao* 普通翠鳥, also *diaoyulang* 釣魚郎; river or common kingfisher); (b) *A. hercules* (*bantou da cuiniao* 斑頭大翠鳥, Blyth's or great blue kingfisher), the status of which seems unclear; (c) *Ceyx erithacus* (*sanzhi cuiniao* 三趾翠鳥, three-toed or Oriental dwarf kingfisher); (d) *Halcyon pileata* (*lan feicui* 藍翡翠 or *heitou feicui* 黑頭翡翠; usually black-capped kingfisher); (e) *H. smyrnensis* (usually *baixiong feicui* 白胸翡翠, white-breasted or white-throated kingfisher); (f) *Ceryle rudis* (*ban yugou* 斑魚狗, [lesser] pied kingfisher); (g) *Megaceryle* or *C. lugubris* (*guan yugou* 冠魚狗, greater [pied] or crested kingfisher). Most of these species have some blue feathers, but the last two are white, with black and other dark parts, especially on their wings.

In former times, it was the blue feathers which sold well on China's markets. This should make (a), (d) and (e) the most likely candidates for the *feicui* in *QTZ*. The colours of (f) and (g) are different, as was just said. That may also apply to (c); its "blue" parts are very dark, almost black, and not as bright as the wings of the other candidates. Moreover, (c) is a small bird, so one may perhaps associate it with the classification *shuicui*.

However, the *QTZ* gives no further specifications, so we cannot decide to which species the text refers. There are additional names for these birds and such current popular terms as *cuiniao* 翠鳥 and *yugou* 魚狗 are largely interchangeable. Moreover, recent ornithological works often provide a different taxonomy; this includes divergent views on local subspecies.[328]

---

326 *Qin jing*, 6a (p. 682).

327 Swinhoe, "On the Ornithology of Hainan", pp. 92–93.

328 Some modern works on kingfishers in Ptak, *Exotische Vögel*. Only one representative work is listed here: Fry et al., *Kingfishers, Bee-eaters and Rollers. A Handbook.* – For kingfishers on Hainan, see, for example, *Hainan dao de niao shou*, pp. 146–151; Shi Haitao et al., *Hainan luqi beizhui dongwu jiansuo*, pp. 139–140.

(43) *Lusi* / Egrets

鷺鷥：有黃、黑觜二種。黃者白露後方至。大小英人，媒致設扣捕之，貨諸市。黑者經年不歸，警覺難捕。

Translation: There are two kinds: [one with] a yellow, [the other with] a black bill. The yellow ones come after the *bailu* 白露 season. Courageous young and grown-up men will [use] lures to catch them and [then] sell them on the markets. The black ones do not return for the whole year; they are very alert and difficult to catch.

Comment: *Bailu*, literally "white dew", designates a solar term and roughly corresponds to the middle of September. The name is similar to the expression *bailu* 白鷺, still used today for *Egretta garzetta*, i.e., the little egret (also *xiao bailu* 小白鷺), a bird found in much of southern China and on Hainan. According to tradition the *bailu* comes when the "dews are thick", hence its name. Egrets have also entered literary works, where one finds them in different contexts.[329]

The passage beginning with "Courageous..." is significantly different from the modern edition of *QTZ* in terms of punctuation marks. We did not find that passage in other Hainan chronicles, but it seems to confirm the common view that the meat of egrets and cranes was a local delicacy and therefore sold on the markets. The two sentences starting with "The yellow ones..." and "The black ones..." are included in later Hainan chronicles.[330] A late Ding'an gazetteer distinguishes between three kinds: a yellow-billed and black-billed variety, and the *bailu*, which it describes as a purely white bird often seen in fields. Another work says the *lusi* birds are completely white. The Wenchang chronicle of the Xianfeng period has an entry on *lu* (without *si*): these birds, we read, are as "white as snow", their crown has ten "long feathers"; the bill and feet are black; when resting, eating and walking, the crowd follows a certain order; it is also called *chongchu* 舂鋤, a name derived from the movements of its head.[331]

The expression *lu* (without *si*) already appears in the *Shi jing*.[332] In the course of time, several combinations involving this character came into being. They usu-

---

329 For general explanations: Read, *Chinese Materia Medica: Avian Drugs*, pp. 21–22 (no. 262). For literary works: Kroll, "The Egret in Medieval Chinese Literature", especially p. 188.

330 Examples: *(Kangxi) Wenchang xianzhi*, j. 9, p. 212; *(Guangxu) Chengmai xianzhi*, j. 1, p. 81.

331 *(Guangxu) Ding'an xianzhi*, j. 1, p. 133; *(Daoguang) Wanzhou zhi*, j. 3, p. 295; *(Xianfeng) Wenchang xianzhi*, j. 2, p. 76. These descriptive elements appear in many earlier sources, for example, in *Piya*, II, j. 7, 15a.

332 See Legge, *The Chinese Classics*. Vol. 4: *The She King*, pp. 205–206, 585, 614–615.

ally refer to egrets and similar birds. These belong to the *Ardeidae* family (also see entry no. 33). The further taxonomy is very complex and has undergone various adjustments. Recent zoological works record several birds under this category for Hainan.[333]

Besides *E. garzetta*, this includes *E. Eulophotes* (*huangzui bailu* 黃嘴白鷺, Chinese egret), *E. alba* (*da bailu* 大白鷺, large or great egret), *E. sacra* (*yanlu* 巖鷺, [eastern] reef heron), *E. intermedia* (*zhong bailu* 中白鷺, intermediate egret), *Bubulcus ibis* (also *B. coromandus*; *niubei lu* 牛背鷺, [eastern] cattle egret), *Ardea cinerea* (*canglu* 蒼鷺, grey heron) and *A. purpurea* (*caolu* 草鷺, purple heron), as well as other birds. They vary in size; some have a black or dark bill, for example *E. garzetta* and *E. alba*, one of the largest members of that group, whereas *E. eulophotes* and certain others have a yellow bill.

Of these birds *E. garzetta* should be the best candidate for the black-billed kind listed in *QTZ*. Until today one of its popular names is *lusi*, and it is among the birds that remain on Hainan during all four seasons. The status of *E. sacra* and *E. intermediata* is less clear, *E. alba* is a migratory bird, so is *E. eulophotes*, which could stand for the yellow-billed variety.

This conclusion may not be in line with the arrangement offered in the Ding'an chronicle. However, it seems compatible with the description in the Wenchang gazetteer: the colours and long hindhead feathers (there should be two shafts) clearly point to *E. garzetta*; *E. alba* does not have such feathers; *E. eulophotes* has them, but, as was said, the bill is different.

Finally, one should exclude *Ardeola bacchus* (*chilu* 池鷺, the Chinese pond heron) from the list of candidates. We had encountered that bird above, in entry no. 33. Its colours are different and the alternative name (*tian*) *niunu*, mentioned in *QTZ*, makes it very improbable that the form *lusi* included this bird as well.

(44) *Qingzhuang* / Grey Herons

青樁: 似鷺而大，亦有白者。集則羣鷺避之。

Translation: It resembles the *lu*, but is larger; there are also white ones. When they flock together, the *lu* groups will avoid them.

---

333 See, for example, *Hainan dao de niao shou*, pp. 37–47; Shi Haitao et al., *Hainan luqi beizhui dongwu jiansuo*, pp. 108–110. For a survey on the *Ardeidae* in China, see, for example, Zhu Xi 朱曦 and Zou Xiaoping 鄒小平, *Zhongguo lulei* 中國鷺類, especially chapter 1. One bird, *Gorsachius magnificus* (see entry no. 37), endemic to Hainan and parts of continental southern China, has received special attention in the literature. See, for example, Lei Fumin and Lu Taichun, *Zhongguo niaolei teyou zhong*, pp. 26–30.

Comment: Identical or partly identical descriptions are found in several Hainan and Guangdong chronicles.[334] Later works also list a bird called *qingzhuang* (a) 青粧, saying it would not be different from the *baihe* 白鶴 ("white crane"), but would have a light blue / green (*qing* 青) plumage. According to a comment found in the Daoguang version of *Guangdong tongzhi* the *qingzhuang* (a) is also called *qingcang* 青鶬.[335]

Modern works usually equate the name *qingzhuang* with *Ardea cinerea*, (*canglu* 蒼鷺, the grey heron). Other popular names for this bird include tianzhuang 田樁, *qingzhuang* (b) 青莊, *laodeng* 老等 and *changbo laodeng* 長脖老等 (the last two derived from the fact that it is often waiting near the water to catch fish).[336] *A. cinerea*, found on Hainan and in other parts of China, is indeed larger than most *lu* birds, for example *Egretta garzetta* (*bailu*, little egret), but it is not larger than *E. alba* (*da bailu*, large egret); these birds were mentioned above (entries no. 33 and no. 43).[337] To this one may add the following observations: The name *baihe* in traditional texts remains vague (it can refer to *E. garzetta* just as *da baihe* 大白鶴 can stand for *E. alba*) and the colour element *qing* should not be taken in a narrow sense, because the dominant colours of *A. cinerea* are white and grey.

Here we may summarize the findings offered in connection with the *QTZ's* entries on cranes / egrets. (1) *Qingzhuang* should represent *A. cinerea*. There are several options for (2) the *lusi* with a black and (3) the *lusi* with a yellow bill; *E. garzetta* is the best candidate for the first kind, *E. eulophotes* for the second variety; but other species may have been meant as well. (4) *Suoyihe* and *niunu* could stand for *Ardeola bacchus*, the Chinese pond heron. If indeed so, then these four would all belong to the *Ardeidae* family. However, it is also possible that the author thought of *Grus grus*, the common crane, when including *suoyihe* / *niunu* in his list.

(45) *Jiling* / Wagtails

鶺鴒: 飛鳴，行搖。

---

334 See, for example, *(Jiajing) Guangdong tongzhi chugao*, j. 31, 18b, and *(Jiajing) Guangdong tongzhi*, II, j. 24, 11b–12a (p. 630).

335 See, for example, *(Daoguang) Qiongzhou fuzhi*, j. 5, 45a (p. 141); *(Daoguang) Guangdong tongzhi*, III, j. 99, p. 251 top.

336 For more traditional names, also see Guo Fu, *Zhongguo gudai dongwuxue shi*, p. 95 no. 22. The first form is rather rare. One example is in *(Xuantong) Ding'an xianzhi*, j. 1, p. 106.

337 See, for example, *Hainan dao de niao shou*, p. 38. Most sources state that *A. cinerea* only spends the winter on Hainan.

Translation: While flying it sings; when walking, [its tail] moves.

Comment: The *jiling* appears in many traditional sources. One of the earliest works is the *Shi jing*, where one also finds the sequence *zai fei zai ming* 載飛載鳴, which could be the origin of the phrase "while flying it sings". There the bird's name occurs as *jiling* 脊令, but the version recorded in *QTZ* is certainly the most common one.[338] Other Hainan and Guangdong chronicles also use this term. Occasionally they add different details: for example, these birds stay on sandy ground and river banks; they are grey and also called *poling* 坡鴒.[339] Today the form 鶺鴒 designates the *Motacillidae* family. Many of its members spend the winter on Hainan and their current English names include the elements "wagtail" or "pipit". We had encountered them above, in entry no. 25.[340]

The brief Chinese text below the name *jiling* is easily found in popular descriptions characterizing the typical behaviour of these birds. This also applies to the tail, which moves up and down, or even to the side, especially when the bird is trying to pick insects off the ground (hence the English name). Given that there are several kinds on Hainan, one should take the form *jiling* as a general term.

(46) *Luci* / **Common Cormorants**

鸕鷀：口中吐雛，善捕魚。其屬鵜鶘、鸂鶒 (?) 俱有。

Translation: It "emits" [its] fledglings from the mouth and is good at catching fish. Pelicans and *xichi* – [both] of its kind – they all are [there].

Comment: The first phrase of this entry also occurs, for example, in an early version of the *Guangdong tongzhi*.[341] Other texts change the wording and replace *tu* 吐 by *tu* 土, which distorts the meaning. One very early text alluding to an "oral birth" is the fragmentary *Yiwu zhi*, quoted in *Taiping yulan* and other texts: "The *luci* does not produce eggs. It conceives / gives birth to the fledglings between ponds and swamps. They also deliver [them] through the mouth, eight to nine at most, five to six at least; they come one after the other, like connected silk balls..."[342]

---

338 See, Legge, *The Chinese Classics*. Vol. 4: *The She King*, pp. 251, 334.

339 See *(Kangxi) Wenchang xianzhi*, j. 9, p. 213, and *(Guangxu) Chengmai xianzhi*, j. 1, p. 80 (same as *QTZ*); *(Jiajing) Guangdong tongzhi chugao*, j. 31, 18b (sandy ground); *(Guangxu) Ding'an xianzhi*, j. 1, p. 132, and *(Xuantong) Ding'an xianzhi*, j. 1, p. 105 (grey, *poling*).

340 See, for example, *Hainan dao de niao shou*, pp. 170–177; Shi Haitao et al., *Hainan luqi beizhui dongwu jiansuo*, pp. 150–151.

341 See *(Jiajing) Guangdong tongzhi chugao*, j. 31, 18b.

342 See *Taiping yulan*, IV, j. 925, 6b (p. 4110); *Piya*, II, j. 6, 19a–b; *Yiyu zhi*, 3a–b.

In traditional texts *luci* is a general term for the cormorant. As is well-known, cormorants can be used to catch fish. The idea of a "mouth birth" should be related to the image of such birds delivering their catch to the fishermen. There are also many stories surrounding these birds.[343]

Today the expression *luci* stands for two things: for the *Phalacrocoracidae* family, i.e., the cormorants, and for *Phalacrocorax carbo* (*putong luci* 普通鸕鷀, common or great cormorant), the only species recorded in the context of Hainan. This bird stays on the island during the winter months and there are several popular names for it, including *hei yulang* 黑魚郎, *yu laoya* 魚老鴉, *wu gui* 烏鬼, *yuying* 魚鷹, etc. These names are self-explicatory; they all derive from the bird's colour and behaviour.[344]

The second sentence in *QTZ*, starting with *tihu* 鵜鶘 (pelicans), is quite modern in the sense that these birds *and* the cormorants both belong to one and the same order, namely the *Pelecaniformes*. Moreover, today the term *tihu* stands for the *Pelecanidae* family of which only one species has been reported for Hainan. This is *Pelecanus philippensis*, or *banzui tihu* 斑嘴鵜鶘, the spotted billed pelican.

Modern works record one additional bird that should be mentioned here: *Sula leucogaster* (*he jianniao* 褐鰹鳥, brown booby). This animal appears on the Xisha Islands 西沙群島 (Paracel Islands) administered by Hainan. It belongs to the *Sulidae* family (again under the *Pelecaniformes*), but its colours are different from those of *Phalacrocorax carbo* and to what extent it also stays on Hainan, remains unclear. Therefore, we may exclude it from our discussion.[345]

As in the case of cormorants, one can encounter many stories on pelicans in traditional works. Pelicans already occur in the *Shi jing* and not infrequently are mentioned after or together with the cormorants.[346]

While the terms *luci* and *tihu* pose no problems – in the Hainan context one should be *Phalacrocorax carbo*, the other *Pelecanus philippensis* – the combination *xichi* raises questions. Both characters are hardly legible in the Tianyige version of *QTZ* (hence the question-mark in our text, above) and it is not clear

---

343 See, for example, Guo Fu, *Zhongguo gudai dongwuxue shi*, pp. 442–444. – Popular and scientific works on cormorants and fishing with cormorants abound. Here are some titles: Laufer, *The Domestication of the Cormorant in China and Japan*; Gudger, "Fishing with the Cormorant"; Manzi and Comes, "Cormorant Fishing in Southwestern China: A Traditional Fishery under Siege"; Zhang Zhongge 張仲葛, "Luci xiaoshi" 鸕鷀小史; Simoons, *Food in China*, pp. 340–342. – An early European description in Du Halde, *Description*, II, p. 168.

344 See *Hainan dao de niao shou*, p. 36.

345 See *Hainan dao de niao shou*, p. 35.

346 Legge, *The Chinese Classics*. Vol. 4: *The She King*, p. 222. Also see, for example, Guo Fu, *Zhongguo gudai dongwuxue shi*, p. 50 no. 119, p. 95 no. 17, p. 113 no. 51, p. 253.

whether their short forms in the modern edition are correct. However, other accounts contain similar combinations many of which are pronounced *xichi* today. Thus, Huang Zuo's *Guangdong tongzhi* has 鸂鶒, while the Daoguang version of that text gives 鸂鷘, to mention just two examples.[347] Besides these expressions one also encounters several "unusal" forms such as *jilai* 鷄瀨.[348]

The relation between these terms and the *yuanyang* 鴛鴦, i.e., the "mandarin duck" (see the next entry here) is a further point of concern. According to one edition of the *Qiongzhou fuzhi* the name *xichi*, in that case 鸂鶒 (or 鶒?, characters not clearly legible), was also used for *yuanyang*, but at the same time this same source carries a separate segment for the former.[349] Other works are less confusing. They make a clear distinction between *yuanyang* and *xichi*, explaining, the latter would occasionally be called *ci yuanyang* 紫鴛鴦, literally "purple mandarin duck" (and not just *yuanyang*), because of its colour (mostly purple, *ci* 紫). In some cases it is also stated that the *xichi* was as large as (or larger than) the mandarin duck.

Here one may turn to the *Bencao gangmu*, cited in the Daoguang text and other works: Li Shizhen makes a distinction between *xichi* and *yuanyang*, he also claims the former would destroy poisonous organisms in its neighbourhood and keep the water clean.[350] The last observation – on the *xichi's* protective properties – is found in many texts, for example Huang Zuo's account.

The earliest reference to that element comes from a fragmentary source, the *Linhai yiwu zhi*, quoted in *Taiping yulan*. Here the animal's name appears as *xichi* 溪鶒. We also learn that the *xichi* was a water bird with "five colours", but this detail and all other features are insufficient for species identification.[351] Read, essentially basing himself on the *Bencao gangmu*, defines it as *Anas fuligula*

---

347 *(Jiajing) Guangdong tongzhi*, II, j. 24, 15b (p. 632), and *(Daoguang) Guangdong tongzhi*, III, j. 99, p. 251 bottom.

348 See *(Wanli) Danzhou zhi,* tianji, p. 36. The entry is quasi the same as the one in *QTZ*. – Note, there are some rhapsodies and old paintings, which provide additional name forms. One painting, attributed to Mao Yi 毛益 (12th century), shows two birds similar to *Nettapus coromandelianus* (*mianfu* 棉凫; cotton teal, cotton pygmy goose). This bird is occasionally associated with Hainan. See, for example, http://news.yidianchina.com/130705/3200.html (July 2014).

349 *(Qianlong) Qiongzhou fuzhi*, j. 1 xia, 97a (p. 103).

350 *Bencao gangmu*, IV, j. 47, pp. 2572–2574, and Read, *Chinese Materia Medica: Avian Drugs*, pp. 20–21 (nos. 259 and 260).

351 *Taiping yulan*, IV, j. 925, 2b (p. 4108). The *Linhai yiwu zhi* is known under different titles. The entry in that source entered several later accounts with no or only very few changes. – Also see, for example, *(Daoguang) Qiongzhou fuzhi*, j. 5, 44a (p. 140), and *(Minguo) Danxian zhi*, j. 3, p. 190. In some of these cases, again, different orthographs are found.

(now usually *Aythya fuligula*), the tufted duck, called *fengtou qianya* 鳳頭潛鴨 in modern zoological works.[352] The head and wings of the male bird show a dark blue-green colour, sometimes with a purple shine, the lower body is mostly white. However, modern zoological works do not record this animal in the context of Hainan.

Of course, there are many other "ducks" on the island, some with a very colourful appearance. Perhaps, then, it is to these birds that the *QTZ* and similar texts refer when mentioning the *xichi* (they do not seem to refer to the *haiya* 海鴨 / domesticated *jiaya* 家鴨 mentioned above, in entry 40). But if so, the brief explanation in our text, namely that the *xichi* would be a *shu* 屬 – a "relative" – of the cormorant, would make little sense. Could this mean that the author had in mind a very different bird, one that had little in common with a "typical" duck, but was similar to a cormorant / pelican?

(47) *Yuanyang* / Mandarin Ducks (?)

鴛鴦：毛羽五色，雌雄相逐濱溪。

Translation: [Their] feathers are in five colours. Female and male follow each other near the shores of creeks.

Comment: The *yuanyang* or mandarin ducks are an emblem of loyalty between husband and wife. They have several popular names (one is *pi niao* 匹鳥: "partner birds"), stand for lasting happiness, are famous for their beautiful plumage and appear in countless stories, poems and popular sayings. Many of these elements are so well-known that they require no comment. Suffice it say that there is an ode in *Shi jing* which bears the title "Yuanyang".[353]

Zoologists normally link the term *yuanyang* to *Aix galericulata* under the *Anatidae* family. But modern ornithological accounts agree in saying that mandarin ducks are not among the natives of Hainan. How, then, should one explain the fact that this bird appears in *QTZ* – and in many other Hainanese chronicles as well, for example in the Wanli and Qianlong editions of *Qiongzhou fuzhi*?[354]

---

352 Read, *Chinese Materia Medica: Avian Drugs*, pp. 20–21 (no. 260).

353 Legge, *The Chinese Classics*. Vol. 4: *The She King*, pp. 387–389; also p. 418, there.

354 *(Wanli) Qiongzhou fuzhi*, j. 3, 99a (p. 76); *(Qianlong) Qiongzhou fuzhi*, j. 1 xia, 97a (p. 103). Both these texts remark that a single duck could not survive without its partner (this element appears in many early works, for example in *Piya*, II, j. 7, 13a). The second text adds, again following earlier sources, that *yuanyang* ducks were named *xichi* 鸂鶒 as well. Also see the previous entry and the previous notes, here. – One or two later works describe the *yuan-*

There are several possible "solutions": (1) *A. galericulata* once belonged to the island's fauna, but later disappeared due to climatic changes or other reasons. (2) *Yuanyang* ducks were confused with a "similar" bird, called *xichi* (see previous entry), as for example in a late version of the *Wanzhou zhi*, where their "anti-poisonous" properties are praised (these properties are normally associated with the *xichi*).[355] (3) Modern zoologists failed to record the presence of *yuanyang* birds on the island; there are recent claims that such ducks were locally sighted in small numbers.[356] (4) Authors coming from the mainland mistook several birds they saw on Hainan – and not only the *xichi* – for mandarin ducks. Indeed, some animals such as *Anas falcata* (*lowen ya* 羅紋鴨, falcated teal), *A. querquedula* (*baimei ya* 白眉鴨, garganey) and *A. formosa* (*hualian ya* 花臉鴨, Baikal teal) are very colourful, just as the male *yuanyang*. Another candidate easily confused with "proper" mandarin ducks may have been *A. crecca*, already mentioned above, under the entry for *shuiya* (see no. 40).

Be this as it may, readers are confronted with the unexpected situation that two "ducks" appear in *QTZ* and other Hainan chronicles (*xichi* and *yuanyang*), none of which belong to the "original" fauna of Hainan. This leaves us with the *shuiya* (besides domesticated ducks) as the only real Hainan (wild) duck, as we saw, although that bird remains equally difficult to identify.

(48) *Geque*

割雀：白晝不見形狀。每歲初春夜，高飛向東。鳴"割割"聲，歲必豐。

Translation: During the day one does not recognize them. Each year, at night in early spring, they fly high [up in the sky], towards the east. [Their] singing sounds [like] "gege" 割割, [which indicates] an opulent year.

Comment: Later Hainan chronicles follow the description found in *QTZ*. But sometimes the bird's name appears as *geque* 閣雀.[357] The *Guangdong tongzhi chugao* calls it *geji* 割鷄. Qu Dajun uses the name *gouge que* 鈎割雀 and also lists a *xuan gouque* 玄鈎雀; the latter comes from Gaoming 高明 (Foshan 佛山

---

yang's plumage in ways that do not match a "standard" mandarin duck: for example, *(Xianfeng) Wenchang xianzhi*, j. 2, p. 78.

355 See *(Daoguang) Wanzhou zhi*, j. 3, p. 295.

356 See, for example, "Hainan bainian niaolei diaocha" 海南百年鳥類調查, based on an article in *Hainan ribao* 海南日報 (18 March 2013). This has led to various reactions, usually in the internet. See, for instance: http://www.hibirds.com/Shownews.asp?ID=354 (August 2014).

357 *(Wanli) Qiongzhou fuzhi*, j. 3, 99a (p. 76); *(Qianlong) Qiongzhou fuzhi*, j. 1 xia, 96b–97a, (pp. 102–103); *(Daoguang) Qiongzhou fuzhi*, j. 5, 44b (p. 140).

area) and sings before the rain. Regarding the voice of the *gouge que*, Qu makes a distinction between an auspicious sound (transcribed *gouge* 鈎割) and a less auspicious one (*gougouge* 鈎鈎割).[358] Evidently such distinctions have led to confusion: the Daoguang version of *Guangdong tongzhi* introduces two birds: *gouge que* 鈎割雀 and *geque* 閣雀; it then adds, both "are more or less the same".[359] Other sources provide similar details, but this mostly comes out of later material and adds little that may help us to identify the bird(s) in question.[360]

Nevertheless, two descriptive elements deserve attention when one tries to solve the problem by finding (phonetically) similar names: (1) according to the *QTZ* we are dealing with a nocturnal bird; (2) this bird should not be too large in size (*que*!). If that is acceptable, we may rule out one option that will rapidly come to mind: *guique* 鬼雀. This is a popular term for *Corvus torquatus* (*baijing ya* 白頸鴉, collared crow). Two other options could be more appropriate: *gouge* 鈎鴿 and *guigege* 鬼各哥.[361] Both stand for owls – not only for *Bubo bubo* (on the continent), but perhaps also for smaller birds, which one finds among the *Strigidae* of Hainan (also see entry no. 38, above). The sounds uttered by these nocturnal animals and their hooked beaks further remind of the characters in the names above and could support such a thesis. However, this is definitely not the last word in that matter.

(49) *Ou* / Sea Gulls

鷗: 臆白翅青，隨潮上下。

Translation: [Its] breast is white, the wings are blue / green; it moves up and down with the tides.

Comment: Identical descriptions are found in several chronicles. Other texts specify three kinds of *ou*: a bluish-white one, a small and a large one.[362] They confirm that these birds, also called *hai'ou* 海鷗, would follow the tides and fly

---

358 *(Jiajing) Guangdong tongzhi chugao*, j. 31, 18b; *Guangdong xinyu*, j. 20, p. 521.

359 *(Daoguang) Guangdong tongzhi*, III, j. 99, p. 252 top.

360 For instance, the *Nanyue biji*, j. 8, 9b–10a (two entries), has *yuan gouque* 元鈎雀 in lieu of *xuan gouque*, but otherwise says the same as Qu Daju.

361 See, for example, Read, *Chinese Materia Medica: Avian Drugs*, pp. 69 (no. 301 C: *guique*, etc.), 86–87 (no. 315: owls); *Hainan dao de niao shou*, pp. 211 (*guique*), 134–137 (small owls).

362 For example *(Wanli) Danzhou zhi*, tianji, p. 36; *(Jiaqing) Chengmai xianzhi*, j. 10, p. 444; *(Guangxu) Chengmai xianzhi*, j. 1, p. 81. – Three kinds: *(Jiajing) Guangdong tongzhi chugao*, j. 31, 18a.

across the Zhanghai 漲海 (an old name for the South China Sea); their eggs would be blue / green and sometimes they would reach the shore in groups, thereby indicating changes in wind and weather.[363] The textual source for all this should be the *Nan Yue zhi* 南越志, a fragmentary work quoted, for example, in *Taiping yulan* (which refers to additional texts as well).[364] There one also finds the name *jiang'ou* 江鷗, which is frequently encountered in the literature, just as the name *shui'ou* 水鷗; moreover, occasionally a difference is made between the last expression and the simple form *ou*.[365]

Ou, *hai'ou* and similar combinations are generic terms. They stand for several species of gulls or sea gulls most of which belong to the *Laridae* family, also called *ou* in modern zoology. Ornithological works dealing with Hainan list about ten species under this group. In most cases their present status is not very clear. This cannot be said of *L. canus*, i.e., *hai'ou* or the common gull. Like many other species it has grey wings with some black parts and a white body. *Qing* in the text – for the colour of the wings – makes no sense, unless one sees in this attribute a reference to a general dark / black / grey or at best "bluish" pattern. The conclusion is that in *QTZ* the term *ou* should represent *L. canus* and perhaps several other birds with a similar appearance.[366]

(50) *Hai'e* / Geese (?)

海鵝：蒼色觜尖，似鵝而大。

Translation: [This bird] has a grey colour and a sharp bill; it resembles a domestic goose (*e* 鵝) but is larger.

Comment: Normally, the single character *e* stands for geese. When combined with a second element it acquires a different meaning. The term *tao'e* 淘鵝, for

---

363 For example, *(Jiajing) Guangdong tongzhi*, II, j. 24, 14b–15a (pp. 631–632); *(Daoguang) Guangdong tongzhi*, III, j. 99, p. 251 top. Note, the first source mistakes the part with the eggs. – On the Zhanghai see Ptak, "Zhanghai: Raum und Konzept. Von den Anfängen bis zur Tang-Zeit.

364 See *Taiping yulan*, IV, j. 925, 3a (p. 4109).

365 See, for example, *Guangdong xinyu*, j. 20, p. 530; *(Wanli) Guangdong tongzhi*, j. 59 (section Qiong 3), 27b; *(Daoguang) Guangdong tongzhi*, III, j. 99, p. 251 top (two entries: *ou* and *shui'ou*). – Additional remarks in *Bencao gangmu*, IV, j. 47, p. 2575–2576; Read, *Chinese Materia Medica: Avian Drugs*, p. 23 (no. 263). Among the "varieties" of *hai'ou* one also finds, for example, a *xinfu* 信鳧; see *Sancai saoyi*, j. 9, 64a (p. 96).

366 See, for example, *Hainan dao de niao shou*, pp. 109–113; Shi Haitao et al., *Hainan luqi beizhui dongwu jiansuo*, pp. 129–130.

example, stands for pelicans,[367] while *tian'e* 天鵝 is used for swans. This last combination also appears in several modern names for birds under the genus *Cygnus*.

Although swans are not recorded in modern ornithological works related to Hainan, some traditional chronicles equate the term *tian'e* with the expression *hai'e*, suggesting that such birds were once at home on the island. Examples are mostly found in Qing texts, which also repeat the brief description given in our entry.[368]

But this is not all. The *QTZ* lists the simple term *e* in a very different section, namely under domestic animals. The relevant entry mentions a grey (*cang* 蒼) and a white variety, adding "those with a hanging neck are from Huai" (垂頸者傳自淮).[369] The last part in this sentence could refer to swans imported from the continent at an earlier point in time – and possibly from the area around or south of the Huai River – while the grey and white birds may stand for geese. But this is merely a guess and not certain at all.

Interestingly, the *Yazhou zhi*, dating from a later period, contains a similar entry. It also distinuishes between a grey (*hui* 灰) and a white variety, while those with a "long neck" (長頸) – i.e., not necessarily a hanging or curved one – "are good at singing"; moreover, "the larger ones weigh seven to eight pounds (*jin* 斤)".[370] Clearly, such a weight suggests geese rather than swans, but the reference to the bird's voice remains obscure.

The picture provided by these different textual elements leaves room for many interpretations. For instance, what should one do with the sharp bill? In the end, one cannot tell what the grey *hai'e* really stood for – whether we are looking at a swan, a goose, or a "related" creature. However, we should probably exclude domestic animals and, more generally, ducks. Above we had encountered other names for birds under the *Anatidae* family and if a duck would have been meant here, the text would certainly offer some comparison, which is not the case.

(51) *Hainiao* / "Sea Birds"

海鳥: 夏秋之夜，驚鳴飛入黎山，土人以爲颶信。

---

367 See, for example, *Yazhou zhi*, j. 4, p. 80.
368 See *(Jiaqing) Chengmai xianzhi*, j. 10, p. 444, and *(Guangxu) Chengmai xianzhi*, j. 1, p. 81; *(Qianlong) Qiongzhou fuzhi*, j. 1 xia, 97a (p. 103), and *(Daoguang) Qiongzhou fuzhi*, j. 5, 44a (p. 140).
369 *QTZ*, j. 9, 1b. Same in *(Wanli) Qiongzhou fuzhi*, j. 3, 98a (p. 76).
370 See, for example, *Yazhou zhi*, j. 4, p. 83.

Translation: In summer and spring nights, [this bird] makes alarming sounds and flies into the Li Mountains 黎山. Local people take this as a [warning] signal for a cyclone.

Comment: Other chronicles contain identical or very similar descriptions, but now and then "Li Mountains" is replaced by the unspecific expression *shan* 山, without further geographical indications.[371] Besides *hainiao* one also finds the names *zhifeng* 知風 (literally: knowing the wind) and *haique* 海雀 in other texts. They appear where the form *hainiao* should occur.[372]

Today the term *haique* represents the *Alcidae*, but there are no members of this family in or around Hainan. The only species in the vicinity of that island is *Synthliboramphus antiquus* (*pianzui haique* 扁嘴海雀; short-billed guillemot, ancient murrelet), a small bird which looks like a mini-penguin; it spends the winter along the shores of Guangdong and Taiwan. This seems to exclude the possibility of equating the name *hainiao* with *S. antiquus*. Since Hainan chronicles provide no further descriptive element in association with *hainiao / zhifeng / haique*, this animal remains unidentified.

(52) *Shuiying* / Osprey (?)

水鷹：似鷹，毛白，觜黃。高飛察魚，墜捕之。

Translation: [This bird] resembles a hawk (*ying* 鷹); the feathers are white, the mouth is yellow. It flies high up to observe the fish [and suddenly] descends to seize them.

Comment: Other local gazetteers refer to the *shuiying* in a similar manner or list this name without explanation.[373] Additional details are rarely provided. One exception is the *Ding'an xianzhi* of the Guangxu period, which says the *shuiying* would also be called *zhi* 鷙.[374]

---

371 See, for example, *(Jiajing) Guangdong tongzhi chugao*, j. 31, 18b; *(Daoguang) Guangdong tongzhi*, III, j. 99, p. 251 top; *(Qianlong) Qiongzhou fuzhi*, j. 1 xia, 96b (p. 102); *(Daoguang) Qiongzhou fuzhi*, j. 5; 44b (p. 140); *(Kangxi) Lin'gao xianzhi*, j. 2, p. 53; *(Qianlong) Qiongshan xianzhi*, j. 9, 26a (p. 542); *(Qianlong) Yazhou zhi*, j. 8, p. 299.

372 *(Kangxi) Wenchang xianzhi*, j. 9, p. 213 (*zhifeng*); *(Wanli) Danzhou zhi,* tianji, p. 36 (*haique*).

373 See, for example, *(Jiajing) Guangdong tongzhi chugao*, j. 31, 18a; *(Jiaqing) Chengmai xianzhi*, j. 10, p. 444; *(Guangxu) Ding'an xianzhi*, j. 1, p. 133; *(Kangxi) Qiongshan xianzhi*, j. 9, 26a (p. 542).

374 See *(Guangxu) Ding'an xianzhi*, j. 1, p. 133.

Readers familiar with the ancient classics may be tempted to link the term *shuiying* to the form *yuying* 魚鷹, because this second combination has been used by early commentators to explain the famous name *jujiu* 雎鳩 (also see entry no. 19, above) in the first ode of *Shi jing*. Another term for *jujiu* is the version *wangju* 王雎, mentioned in the *Erya*. Some sources also provide the name *e* 鶚 next to the form *yuying*. English translations of the *Shi jing* normally give "osprey" for *jujiu*. Hence, the above can be reduced to the following equation: *jujiu* (*Shi jing*) = *wangju* = *yuying* = *e* = osprey. Most authors have accepted this "solution", but others believe the general tenor and background of the *Shi jing* ode may not be compatible with a ferocious bird of prey; therefore, the name *jujiu* should point to a "simple" duck or another water bird.[375] However, these details are of no concern to us, because the *QTZ* clearly refers to a bird of prey.

Here, then, we should better turn to later sources, especially to Li Shizhen's entry on the *e* bird, which offers additional explanations. This is what Li has to say: The *e* 鶚 has a startling appearance (*e* 愕), hence its name. The element *ju* 雎 has to do with its excellent eyesight. Another term – *xiaku wu* 下窟烏 – mirrors its ability to collect food from caves. When flying over water, it fans the surface with its wings, thus forcing the fish to come up; this explains the name *feibo* 沸波. Some birds have a white tail; these are the *baijue* 白鷢. Generally however, the *e* has a khaki colour and deep-set eyes. Male and female birds get along well with each other – evidently a concession to the ode in *Shi jing*. During the mating season they fly in pairs. They mostly feed on fish, therefore people in Jiangbiao 江表 (south side of the Yangzi River) call them *shiyu ying* 食魚鷹; but they also eat snakes.[376]

These elements do point to the osprey or a similar bird. Modern ornithology distinguishes three kinds of such birds found on Hainan. They belong to the *Accipitridae* family (also see entries 12 and 14 above) and bear the following names: *Ichthyophaga nana* (or *I. humilis*; *yudiao* 漁鵰 or 雕; [lesser] fish eagle), *Haliaeetus leucogaster* (*baifu haidiao* 白腹海鵰 or 雕; white-bellied sea eagle), and *Pandion haliaetus* (*e* 鶚, *yuying* 魚鷹, etc.; osprey).[377] The status of the first two is uncertain. *P. haliaetus* seems rare and was mostly recorded along the eastern side of Hainan. In theory, the term *shuiying* in *QTZ* could stand for any of

---

375 See Legge, *The Chinese Classics*. Vol. 4: *The She King*, pp. 1–3; *Erya zhushu*, j. 17, p. 306; *Qin jing*, 4a (p. 681); *Erya yi*, II, j. 16, pp. 170–171. – More details in Ptak, *Birds and Beasts in Chinese Texts and Trade*, pp. 13–15.

376 See *Bencao gangmu*, IV, j. 49, pp. 2673–2674; Read, *Chinese Materia Medica: Avian Drugs*, pp. 84–85 (no. 313). – As usual, in giving these details, Li Shizhen cites from many earlier sources.

377 See, *Hainan dao de niao shou*, pp. 61, 64–65; Shi Haitao et al., *Hainan luqi jizhui dongwu jiansuo*, pp. 113 and 116.

these species, however their beaks are not yellow. If the white body colour is a criterion, preference should be given to the last kind.

Concluding Remarks

The bird list in *QTZ* is the earliest extant catalogue describing the avian world of Hainan. Therefore this list should be of interest to scholars working on the history of ornithology. But the above also shows that translating such texts raises many questions. Long comments are necessary to disentangle the meaning of certain passages and, above all, to identify bird names. Each comment could be extended to a special study in order to find out more about the background of individual expressions, the stories associated with them, and the quotations drawn from earlier material.

At present certain issues cannot be solved in a satisfactory way; this concerns, for example, the entries on the *geque* and *limu que* birds which require additional research. Notwithstanding, the above allows us to offer some general conclusions in regard to names. For instance, many names are generic terms used for a set of similar or seemingly related animals; other expressions designate one particular species. By and large the terminology follows older traditions developed on the continent, especially in the areas around the Yangzi or even farther north. Such terms were gradually imported to Hainan in earlier times, prior to the compilation of our text; presumably it was scholar-officials and migrants from other parts of China who circulated the imported expressions. Now and then this led to distortions because certain terms came to be used for different animals.

Other observations concern the categorization of birds into subgroups. Such terms as *ji* 鷄 and *que* 雀 pose enormous problems, as we have seen, because one cannot always bring them in line with the current scientific taxonomy. In other cases traditional Chinese concepts, implicitly present in short descriptions or the formal arrangement of a text, are not too far from our modern views. The close association of cormorants and pelicans is one example. Furthermore, there are generic names under which one can identify further categories with several kinds. This suggests that scholars had thought about a hierarchical organisation of the avian world. Such arrangements may or may not be different from the ones in continental bird lists of the Ming period.

The comments presented after each entry, we must admit, are weak in two regards: they neglect the issue of local dialects and languages. Transcribing bird names according to Hainanese standards, the Cantonese pronunciation or even

Fujianese forms might shed further light on open questions. The second theme concerns the distribution of species on the island itself. Local chronicles, especially later works on single districts, may tell us something about regional variations. The *QTZ* is of course different in that regard; it represents Hainan in its totality and not just one of its natural subsystems.

Several entries suggest that people on Hainan were aware of minor differences between the continental fauna and the local animal world. But these differences remain unexplained. From the text it also transpires that officials and / or ordinary migrants took a small number of birds to the island, where these animals began to multiply. Whether local chronicles permit us to also conclude that in other cases certain species disappeared from Hainan in the course of the Ming and Qing dynasties, awaits a thorough discussion.

Occidental scholars have argued that learned Chinese rarely undertook serious efforts to define differentiae through systematic species analysis. The many names and terms one finds in old sources, from very early times through to later periods, would merely reflect a continued interest in lexicography, but not in the fauna as such; furthermore, there would be a tendency to see the animal world as one element within a greater cosmological setting, and not as a separate entity. Yes, this is partly true; the idea of metamorphosis, alluded to in *QTZ*, is an indicator for these dimensions. But stating that men in Zhou and Han times, and in later periods, had no eye for simple differentiae goes much too far. The descriptions and comparisons associated with animal names, the way these terms were grouped, re-grouped and defined in early texts, clearly suggests that ancient observers had thought of criteria suitable for a better understanding of the world around them.

A related theme is the internal arrangement of the animal lists that one finds in traditional accounts. Some older works start with the *fenghuang* 鳳凰; the *he* 鶴, *yingwu* 鸚鵡 and other "noble" birds appear next, while "ordinary" or "minor" animals only come towards the end of a sequence. Ming and Qing gazetteers often deviate from such patterns, but the reasons are not always clear. At times these chronicles also echo older conventions; in works on the southern fauna, for example, one finds several textual clusters that suggest early influence. One such cluster is made up of "talking birds": the *yingwu*, the *quyu* 鴝鵒, and the *qinjiliao* 秦吉了. Other arrangements concern "waterbirds", pheasants and fowl, birds of prey, and so forth. These constructions reflect continued efforts to classify the fauna of a region; in the case of Hainan's chronicles they seem to be carried, in part at least, from one text to the next.

The bird list in *QTZ* only constitutes a small segment within a huge compilation of local data on Hainan. A plain translation of the entire book, without footnotes, would probably give more than one thousand English pages. Such a mass of information suggests that, for reasons of time and economy, the author / compilator was certainly grateful for being able to rely on earlier material for some of the entries included in the avian section of his text; however, several elements in that chapter clearly reflect the circulation of *in situ* knowledge not necessarily available through continental sources. If that is acceptable one may also say that the bird part in *QTZ* tries to combine local observations with conventions ultimately adopted from the mainland, which, once again, presupposes careful deliberation on implicit criteria for differentiation.

In imperial China experienced readers always expected to find citations from well-known sources in official works such as local gazetteers. This would add to the weight and reliability of a text and expose the scholarship of its author. The repeated references to Su Shi are an expression of these traditions. The same applies to the quotations from the chronicle of Wang Zuo, a leading Hainan scholar under the Ming. One may add that Tang Zhou was one of Wang's students.

Occasionally, Ming and Qing works with notes on animals and plants allude to strange legends and fantastic elements. European sources of the medieval and early modern periods also refer to *mirabilia*. Even the Jesuits, who came to China in the sixteenth and seventeenth century, followed these traditions. However, in the course of time the "unreal" becomes less important. The bird lists in *QTZ* and many later Hainan gazetteers are very realistic and not overloaded with invented stories. For instance, as was said, there is no chapter on the *fenghuang*; references to this fantastic creature are banned to the entry on the *wuse que*.

Besides providing remarks on the colour, shape, behaviour and other characteristics of animals, Chinese gazetteers also mention the uses of animal products in medicine and daily life. The birds' nests in *QTZ* constitute one example. But European-language studies on animal products in the context of early modern China and its neighbours are still rare. A detailed analysis of Qing chronicles in particular should bring to light many unknown facts regarding these dimensions of the past.

In sum, there are various reasons to study references to animals in imperial China. Ecological, zoological, sociological and other considerations may encourage research in that direction. Presumably the fauna of several regions underwent dramatic changes during the last few centuries. Certain animals disappeared, others "moved in"; this should also apply to Hainan. Perhaps, then, the present booklet will be of some use for the further investigation of such issues.

Bibliography

Abbreviations

BBCS = Baibu congshu jicheng 百部叢書集成 (Taiwan Yiwen yinshuguan).

HNDFZCK = Hainan difangzhi congkan 海南地方志叢刊, ed. by Hainan difang wenxian congshu bianzuan weiyuanhui 海南地方文獻叢書編纂委員會 (Hong Shouxiang 洪壽祥, Zhou Weimin 周偉民 et al., eds.) 68 vols. (Haikou: Hainan chubanshe, 2003–2006).

QTZ = (Zhengde) Qiongtai zhi, Tianyi ge cang Mingdai fangzhi xuankan 天一閣藏明代方志選刊 (Shanghai guji shudian).

SKQS = Jingyin Wenyuange Siku quanshu 景印文淵閣四庫全書 (Taiwan Shangwu yinshuguan).

Chinese "Primary" Sources

Note: (1) All works are listed by titles. (2) This includes the dynastic annals, local chronicles and other texts, as well as annotated editions, complete works and anthologies, but not translations into European languages, which appear in the next section. (3) Congshu and series titles are given in abbreviated form; for collections that are not so common, we added publication details. (4) Chronicles of individual Hainan counties: we mostly relied on the HNDFZCK edition, but there are certain exceptions. For works on the entire island and Guangdong, we used older material.

Anchun pu 鵪鶉譜, by Cheng Shilin 程石鄰 (Congshu jicheng xubian ed.).

Bamin tongzhi 八閩通志, by Huang Zhongzhao 黃仲昭, 2 vols. (Fuzhou: Fujian renmin chubanshe, 2006).

Bencao gangmu 本草綱目, by Li Shizhen 李時珍, 4 vols. (Beijing: Renmin weisheng chubanshe, 1975–1981).

Bencao jing 本草經, ed. by Cao Yuanyu 曹元宇 (Shanghai: Shanghai kexue jishu chubanshe, 1987).

Bencao yan yi 本草衍義 by Kou Zongshi 寇宗奭, ed. by Yan Zhenghua 顏正華, Chang Zhangfu 常章 and Huang Youqun 黃幼群 (Beijing: Renmin weisheng chubanshe, 1990).

Cao Zhi ji jiaozhu 曹植集校注, by Cao Zhi 曹植, ed. by Zhao Youwen 趙幼文 (Beijing: Renmin wenxue chubanshe, 1984).

(Kangxi) Changhua xianzhi （康熙）昌化縣志, by Fang Dai 方岱, revised by Qu Zhican 瞿之璨, ed. by Feng Junhua 馮俊華, HNDFZCK (Haikou: Hainan chubanshe, 2004).

(Jiaqing) Chenghai xianzhi （嘉慶）澄海縣志, by Li Shuji 李書吉, ser. Zhongguo fangzhi congshu, Hua'nan 62 (Taibei: Chengwen chubanshe, 1967).

(Guangxu) Chengmai xianzhi（光緒）澄邁縣志, by Long Chaoyi 龍朝翊, comp. by Chen Suoneng 陳所能 et al., ed. by Chen Hongmai 陳鴻邁, HNDFZCK (Haikou: Hainan chubanshe, 2004).

(Jiaqing) Chengmai xianzhi（嘉慶）澄邁縣志, by Xie Jishao 謝濟韶, comp. by Li Guangxian 李光先, ed. by Chen Hongmai 陳鴻邁, HNDFZCK (Haikou: Hainan chubanshe, 2004).

Chongbian Qiongtai gao 重編瓊臺藁, by Qiu Jun 丘濬, ed. by Qiu Ergu 丘爾穀 (SKQS ed. 1248).

Chongji Bencao shiyi 重輯本草拾遺, by Chen Cangqi 陳藏器, revised by Na Yi 那琦 (Taizhong: Huaxia wenxian ziliao chubanshe, 1988).

Da Dai Li ji buzhu (fu jiaozheng Kong shi Da Dai Li ji buzhu) 大戴禮記補注（附校正孔氏大戴禮記補注）, by Kong Guangsen 孔廣森, ed. by Wang Fengxian 王豐先, ser. Shisan jing Qing ren zhushu (Beijing: Zhonghua shuju, 2013).

Da Ming yitong zhi 大明一統志, by Li Xian 李賢 et al., 10 vols. (Taibei: Wenhai chubanshe, 1965).

Dade Nanhai zhi canben 大德南海志殘本, by Chen Dazhen 陳大震 (Guangzhou: Guangzhou shi difangzhi yanjiusuo, 1986).

(Minguo) Danxian zhi（民國）儋縣志, by Peng Yuanzao 彭元藻 and Zeng Youwen 曾友文, comp. by Wang Guoxian 王國憲, ed. by Lin Guanqun 林冠群, HNDFZCK (Haikou: Hainan chubanshe, 2003).

(Kangxi) Danzhou zhi（康熙）儋州志, by Han Youzhong 韓祐重, ed. by Lin Guanqun 林冠群, HNDFZCK (Haikou: Hainan chubanshe, 2004).

(Wanli) Danzhou zhi（萬曆）儋州志, by Zeng Bangtai 曾邦泰 et al., ed. by Lin Guanqun 林冠群, HNDFZCK (Haikou: Hainan chubanshe, 2004).

(Guangxu) Ding'an xianzhi（光緒）定安縣志, by Wu Yinglian 吳應廉, comp. by Wang Yangdou 王映斗, ed. by Zheng Xingshun 鄭行順 and Chen Jianguo 陳建國, HNDFZCK, 2 vols. (Haikou: Hainan chubabshe, 2003).

(Xuantong) Ding'an xianzhi（宣統）定安縣志, continued by 宋席珍, ed. by Zheng Xingshun 鄭行順 and Chen Jianguo 陳建國, HNDFZCK, 2 vols. (Haikou: Hainan chubanshe, 2003).

Erya yi 爾雅翼, by Luo Yuan 羅願 and Hong Yanzu 洪焱祖, 4 vols. (Congshu jicheng chubian ed.).

Erya yishu 爾雅義疏, by Hao Yixing's 郝懿行, 8 vols. (ed. with preface of 1856, with the name of Hu Tingshi 胡珽識).

Erya zhushu 爾雅註疏, comm. by Guo Pu 郭璞 and Xing Bing 邢昺, ed. by Li Chuanshu 李傳書 and Xu Chaohua 徐朝華 (Beijing: Beijing daxue chubanshe, 1999).

Fangyu shenglan 方輿勝覽, by Zhu Mu 祝穆, suppl. by Zhu Zhu 祝洙, ed. by Shi Hejin 施和金, ser. Zhongguo gudai dili zongzhi congkan, 3 vols. (Beijing: Zhonghua shuju, 2003).

(Minguo) Gan'en xianzhi（民國）感恩縣志, by Zhou Wenhai 周文海, comp. by Lu Zongtang 盧宗棠 and Tang Zhiying 唐之瑩 (comp.), ed. by Du Huizhen 杜惠珍 and Cai Changqi 蔡昌其, HNDFZCK (Haikou: Hainan chubanshe, 2004).

Ge zhi jing yuan 格致鏡原, by Chen Yuanlong 陳元龍, 16 vols. without numbers (ed. of 1888, no specification).

(Daoguang) Guangdong tongzhi（道光）廣東通志, by Ruan Yuan 阮元, ed. by Chen Changqi 陳昌齊, ser. Xuxiu siku quanshu 669–675, 7 vols. (Shanghai: Shanghai guji chubanshe, 1995).

(Jiajing) Guangdong tongzhi（嘉靖）廣東通志, by Huang Zuo 黃佐, 4 vols. (Hong Kong: Dadong tushu gongsi, 1977).

(Wanli) Guangdong tongzhi（萬曆）廣東通志, by Guo Fei's 郭棐, several prefaces (rare edition).

(Jiajing) Guangdong tongzhi chugao（嘉靖）廣東通志初稿, by Dai Jing 戴璟, comp. by Zhang Yue 張岳, ser. Beijing tushuguan guji zhenben congkan 38 (Beijing: Shumu wenxian chubanshe, 1988).

Guangdong xinyu 廣東新語, by Qu Dajun 屈大均 (Beijing: Zhonghua shuju, 1975).

Guangya shuzheng 廣雅疏証, by Zhang Yi 張揖, comm. by Wang Niansun 王念孫, ed. by Zhong Zishun 鐘字訊 (Beijing: Zhonghua shuju, 1983).

Gugong niao pu 故宮鳥譜, ed. by Qin Xiaoyi 秦孝儀 et al., 4 vols. (Taibei: Guoli gugong bowuguan, 1999).

Guihai yuheng zhi jiyi jiaozhu 桂海虞衡志輯佚校注, by Fan Chengda 范成大, ed. by Hu Qiwang 胡起望 and Tan Guangguang 覃光廣 (Chengdu: Sichuan minzu chubanshe, 1986).

Gujin zhu 古今注, by Cui Bao 崔豹 (Congshu jicheng chubian ed.).

Hai yu 海語, by Huang Zhong 黄衷, Lingnan yishu (BBCS ed.).

Haicha yulu 海槎餘錄, by Gu Jie 顧玠, Jilu huibian (BBCS ed.).

Han shu 漢書, by Ban Guo 班固, 12 vols. (Beijing: Zhonghua shuju, 1987).

Han Wei liuchao baisan jia ji 漢魏六朝百三家集 (SKQS ed. 1412–1416).

Huainanzi jishi 淮南子集釋, ed. by He Ning 何寧, 3 vols. (Beijing: Zhonghua shuju, 1998).

Huang shi ri chao 黃式日鈔, by Huang Zhen 黃震 (SKQS ed. 707–708).

Ji pu jiaoshi – douji de siyang guanli 鷄譜校釋 – 鬥鷄的飼養管理, ed. by Wang Zichun 汪子春 (Beijing: Nongye chubanshe, 1989).

Jiannan shigao 劍南詩稾, by Lu You 陸游, 2 vols. (SKQS ed. 1162–1163).

Jilei ji 鷄肋集, by Wang Zuo 王佐 (Wang Tongxiang 王桐鄉), ed. by Liu Jiansan 劉劍三 (Haikou: Hainan chubanshe, 2004).

Jin shu 晉書, by Fang Xuanling 房玄齡, 10 vols. (Beijing: Zhonghua shuju, 1974).

Jiu Tang shu 舊唐書, by Liu Xun 劉昫, 16 vols. (Beijing: Zhonghua shuju, 1975).

Kaiyuan Tianbao yishi shi zhong 開元天寶遺事十種, by Wang Renyu 王仁裕, ed. by Ding Ruming 丁如明 (Shanghai: Shanghai guji chubanshe, 1985).

Li Bai ji jiaozhu 李白集校注, by Li Bai 李白, ed. by Qu Tuiyuan 瞿蛻園 and Zhu Jincheng 朱金城, 4 vols. (Shanghai: Shanghai guji chubanshe, 1980).

Li ji jin zhu jin yi 禮記今註今譯, comm. and tr. by Wang Meng'ou 王夢鷗, ed. by Wang Yunwu 王雲五, 2 vols. (rev. ed. Taibei: Taiwan Shangwu yinshuguan, 1984).

Liezi jishi 列子集釋, by Yang Bojun 楊伯俊, Xinbian Zhuzi jicheng, ser. I (Beijing: Zhonghua shuju, 1979).

(Kangxi) Lin'gao xianzhi（康熙）臨高縣志, by Fan Shu 樊庶, ed. by Liu Jiansan 劉劍三 and Zheng Xingshun 鄭行順, HNDFZCK (Haikou: Hainan chubanshe, 2004).

Lingbiao lu yi 嶺表錄異, by Liu Xun 劉恂, Juzhenban congshu (BBCS ed.).

(Kangxi) Lingshui xianzhi（康熙）陵水縣志, by Pan Tinghou 潘廷侯, ed. by Zheng Xingshun 鄭行順, HNDFZCK (Haikou: Hainan chubanshe, 2004).

(Qianlong) Lingshui xianzhi（乾隆）陵水縣志, by Ju Yunkui 瞿雲魁, ed. by Zheng Xingshun 鄭行順, HNDFZCK (Haikou: Hainan chubanshe, 2004).

Lingwai daida jiaozhu 嶺外代答校注, by Zhou Qufei 周去非, ed. by Yang Wuquan 楊武泉, ser. Zhongwai jiaotong shiji congkan (Beijing: Zhonghua shuju, 1999).

Liu Zongyuan quanji 柳宗元全集, ed. by Cao Minggang 曹明綱 (Shanghai: Shanghai guji chubanshe, 1997).

Luofushan zhi huibian 羅浮山志會編, by Song Guangye 宋廣業, in *Zangwai daoshu* 藏外道書, vol. 19 (Chengdu: Bashu shushe, 1994).

Mengxi bitan jiaozheng 夢溪筆談校證, by Shen Gua 沈括, ed. by Hu Daojing 胡道靜, 2 vols. (Shanghai: Shanghai guji chubanshe, 1987).

Ming shi 明史, by Zhang Tingyu 張廷玉 et al., 28 vols. (Beijing: Zhonghua shuju, 1974).

Ming shi jishi benmo 明史紀事本末, by Gu Yingtai 谷應泰, 4 vols. (Beijing: Zhonghua shuju, 1977).

Mingdai Ge jing Qingdai Ge pu 明代鴿經 清宮鴿譜, ed. by Wang Shixiang 王世襄 and Zhao Chuanji 趙傳集 (Beijing: Shenghuo dushu xinzhi sanlian shudian, 2013).

Nanyue biji 南越筆記, by Li Diaoyuan 李調元, 4 vols., Han hai (BBCS ed.).

Ouyang Xiu quanji 歐陽修全集, by Ouyang Xiu 歐陽修, ed. By Li Yi'an 李逸安, ser. Zhongguo gudian wenxue congshu, 6 vols. (Beijing: Zhonghua shuju, 2001).

Piya 埤雅, by Lu Dian's 陸佃, 5 vols., Wuya quanshu (BBCS ed.).

Qin jing 禽經, by Shi Kuang 師曠 (SKQS ed. 847).

Qing yi lu 清異錄, ba Tao Gu 陶谷, in Tao Zongyi 陶宗儀, *Shuo fu* 說郛, 12 vols. (Beijing: Zhongguo shudian, 1986).

(Kangxi) Qiongshan xianzhi（康熙）瓊山縣志, by Wang Zhi 王贄, comp. by Guan Bideng 關必登, ser. Riben cang Zhongguo hanjian difangzhi congkan (Beijing: Shumu wenxian chubanshe, 1992).

(Xianfeng) Qiongshan xianhi（咸豐）瓊山縣志, by Li Wenheng 李文恒 (?), comp. by Zheng Wencai 鄭文彩, 16 vols. (ed. of 1857).

(Zhengde) Qiongtai zhi（正德）瓊臺志, by Tang Zhou 唐冑, ser. Tianyi ge cang Mingdai fangzhi xuankan, 2 vols. (Shanghai: Shanghai guji shudian, 1964).

(Zhengde) Qiongtai zhi（正德）瓊臺志, by Tang Zhou 唐冑, ed. by Peng Jingzhong 彭靜中, HNDFZCK, 2 vols. (Haikou: Hainan chubanshe, 2006).

(Daoguang) Qiongzhou fuzhi（道光）瓊州府志, by Ming 明誼, comp. by Zhang Yueshong 張岳崧, ed. by Li Lin 李琳, ser. HNDFZCK, 4 vols. (Haikou: Hainan chubanshe, 2006).

(Qianlong) Qiongzhou fuzhi（乾隆）瓊州府志, by Xiao Yingzhi 蕭應植 and Chen Jingxun 陳景塤, ser. Xuxiu siku quanshu 676 (Shanghai: Shanghai guji chubanshe, 2002).

(Wanli) Qiongzhou fuzhi（萬曆）瓊州府志, by Dai Xi 戴熺, comp. by Ouyang Can 歐陽燦, Cai Guangqian 蔡光前 et al., ser. Riben cang Zhongguo hanjian difangzhi congkan (Beijing: Shumu wenxian chubanshe, 1990).

Quan Song shi 全宋詩, ed. by Beijing daxue guwenxian yanjiusuo 北京大學古文獻研究所, 72 vols. (Beijing: Beijing daxue chubanshe, 1991–1998).

Sancai zaoyi 三才藻異, comp. by Tu Cuizhong 屠粹忠 (Siku quanshu cunmu congshu, zi bu 229).

Shanhai jing jiaozhu 山海經校注, ed. by Yuan Ke 袁珂 (Taibei: Liren shuju, 1982).

Shuowen jiezi zhu 說文解字注, by Xu Shen 許慎, comm. by Duan Yucai 段玉裁 (Shanghai: Shanghai guji chubanshe, 1981).

Song shi 宋史, by Tuo Tuo 脫脫 et al., 40 vols. (Beijing: Zhonghua shuju, 1985).

Su Shi quanji 蘇軾全集, ed. by Fu Cheng 傅成 and Mu Chou 穆儔, 3 vols. (Shanghai: Shanghai guji chubanshe, 2000).

Su Shi shiji 蘇軾詩集, comp. by Wang Wengao 王文誥, ed. by Kong Fanli 孔凡禮, 8 vols. (Beijing: Zhonghua shuju, 1982).

Sui shu 隋書, by Wei Zheng 魏徵 et al., 6 vols. (Beijing: Zhonghua shuju, 1973).

Taiping yulan 太平御覽, by Li Fang 李昉 et al., 4 vols. (Beijing: Zhonghua shuju, 1985).

Tang Xianzu shi wen ji 湯顯祖詩文集, ed. by Xu Shuofang 徐朔方, 2 vols. (Shanghai: Shanghai guji chubanshe, 1982).

Tao Yuanming ji jianzhu 陶淵明集箋注, ed. by Yuan Xingpei 袁行霈, ser. Zhongguo gudian wenxue congshu (Beijing: Zhonghua shuju, 2003).

Tong dian 通典, comp. by Du You 杜佑, ed. by Wang Wenjin 王文錦 et al., 5 vols. (Beijing: Zhonghua shuju, 1988).

Tong zhi 通志, by Zheng Qiao 鄭樵, Shi tong, 3 vols. (Hangzhou: Zhejiang guji chubanshe, 1988).

Wanling ji 宛陵集, by Mei Yaochen 梅堯臣 (SKQS ed. 1099).

(Daoguang) Wanzhou zhi（道光）萬州志, by Hu Duanshu 胡端書, comp. by Yang Shijin 楊士錦 and Wu Hongqing 吳鴻清, ed. by Chen Yongzhi 陳智勇 and Wang Ruo 王若, HNDFZCK (Haikou: Hainan chubabshe, 2004).

(Kangxi) Wanzhou zhi（康熙）萬州志, comp. by Li Yan 李琰, ed. by Chen Yongzhi 陳智勇 and Wang Ruo 王若, HNDFZCK (Haikou: Hainan chubanshe, 2004).

Wei Yingwu shiji xinian jiaojian 韋應物詩集系年校箋, by Wei Yingwu 韋應物, ed. by Sun Wang 孫望 (Beijing: Zhonghua shuju, 2002).

(Kangxi) Wenchang xianzhi（康熙）文昌縣志, by Ma Ribing 馬日炳, ed. by Lai Qingshou 賴青壽 and Yan Yanhong 顏艷紅, HNDFZCK (Haikou: Hainan chubabshe, 2003).

(Minguo) Wenchang xianzhi（民國）文昌縣志, drafted by Li Zhongyue 李鍾嶽 et al., comp. by Lin Daiying 林帶英 et al., HNDFZCK, 2 vols. (Haikou: Hainan chubabshe, 2003).

(Xianfeng) Wenchang xianzhi（咸豐）文昌縣志, drafted by Zhang Pei 張霈 et al., comp. by Lin Yandian 林燕典, ed. by Yan Yanhong 顏艷紅 and Lai Qingshou 賴青壽, HNDFZCK, 2 vols. (Haikou: Hainan chubanshe, 2003).

Xiang he jing 相鶴經, attr. to Fu Qiu 浮丘, revised by Wang Anshi 王安石, Wuchao xiaoshuo da guan 9 (Shanghai: Saoye Shanfang, 1926).

Xijing zaji 西京雜記, by Ge Hong 葛洪, in Cheng Yizhong 程毅中 (ed.), *Xijing zaji* and *Yan dan zi* 燕丹子, ser. Gu xiaoshuo congkan (Beijing: Zhonghua shuju, 1985).

Xinxiu bencao: jifu ben 新修本草: 輯復本, by Su Jing 蘇敬, ed. by Shang Zhijun 尚志鈞 (Hefei: Anhui kexue jishu chubanshe, 1981).

Yanzhou xugao 弇州續稿, by Wang Shizhen 王世貞 (SKQS ed. 1282–1284).

Yazhou zhi 崖州志, by Zhang Xi 張巂, by Xing Dinglun 刑定綸 and Zhao Yiqian 趙以謙 (as authors, comp.), ed. by Guo Moruo 郭沫若 (Guangzhou: Guangdong renmin chubanshe, 1983).

(Qianlong) Yazhou zhi（乾隆）崖州志, enlarged by Song Jin 宋錦, by Huang Dehou 黃德厚, ed. by Hou Min 侯敏 and Chen Zhiyong 陳智勇 HNDFZCK (Haikou: Hainan chubanshe, 2006).

Yiwu zhi 異物志, by Yang Xiaoyuan 楊孝元, ed. by Zeng Zhao 曾釗, Lingnan yishu (BBCS ed.).

Yongle dadian fangzhi jiyi 永樂大典方志輯佚, ed. by Ma Rong 馬蓉 et al., 5 vols. (Beijing: Zhonghua shuju, 2004).

Youyang zazu 酉陽雜俎, by Duan Chengshi 段成式, ser. Zhongguo shixue congshu xubian (Taibei: Taiwan xuesheng shuju, 1985).

Yuan shi yiwen zhi 元史藝文志, by Qian Daxin 錢大昕, ed. by Tian Hanyun 田漢雲, in Chen Wenhe 陳文和 (ed.), *Jiading Qian Daxin quanji* 嘉定錢大昕全集, 10 vols. (Nanjing: Jiangsu guji chubanshe, 1997).

Yuchu xinzhi 虞初新志, comp. by Zhang Chao 張潮 (Beijing: Wenxue guji kanxingshe, 1954).

Yudi jisheng 輿地紀勝, by Wang Xiangzhi 王象之, 8 vols. (Beijing: Zhonghua shuju, 1992).

Zhang Jiuling ji jiaozhu 張九齡集校注, by Zhang Jiuling, ed. by Xiong Fei 熊飛, ser. Zhongguo gudian wenxue congshu, 3 vols. (Beijing: Zhonghua shuju, 2008).

Zheng Gu shiji jianzhu 鄭谷詩集箋注, by Zheng Gu 鄭谷, ed. by Yan Shoucheng 嚴壽澂, Huang Ming 黃明 and Zhao Changping 趙昌平 (Shanghai: Shanghai guji chubanshe, 1991).

Zhenla fengtu ji jiaozhu 眞臘風土記校注, by Zhou Daguan, ed. by Jin Ronghua 金榮華 (Taibei: Zhengzhong shuju, 1976).

Zhufan zhi zhubu 諸蕃志注補, ed. by Han Zhenhua 韓振華, in Han Zhenhua zhuzuo zhengli xiaozu 韓振華著作整理小組 (ed.), *Han Zhenhua xuanji* 韓振華選集, vol. 2., ser. Centre of Asian Studies Occasional Papers and Monographs 134.2 (Hong Kong: Centre of Asian Studies, 2000).

"Modern" Secondary Works

This includes bird catalogues, translations of Chinese sources into European languages, all other works in European languages, including two or three old titles, and modern works in Chinese. All works are listed by author, translator or editor. Internet sources are not given; they only appear in the notes.

Allan, Sarah: *The Shape of the Turtle: Myth, Art and Cosmos in Early China* (Albany: State University of New York Press, 1991).

Aomen tebie xingzhengqu, Minzheng zongshu 澳門特別行政區, 民政總署... , i.e., Lei Wai Nong 李偉禮 et al. (eds.): *Aomen niaolei* 澳門鳥類 (*Aves de Macau, Birds of Macau*) (Macau: Aomen tebie..., 2010).

Bielenstein, Hans: *Diplomacy and Trade in the Chinese World, 589–1276* (Leiden, etc.: Brill, 2005).

Bocci, Chiara: "Il leopardo nell'antica Cina fra danze sciamaniche e stendardi", in Ptak (ed.), *Tiere im alten China*, pp. 99–130.

Bu Weibo 補維波: *Tang Zhou yanjiu* 唐胄研究 (Hainan shifan daxue, 2012; unpubl. MA thesis).

Chaves, Jonathan: *Mei Yao-ch'en and the Development of Early Sung Poetry* (New York, etc.: Columbia University Press, 1976).

Chen Jinxjian 陳金現: "Su Shi zai Danzhou de shenfen rentong" 蘇軾在儋州的身份認同, *Guowen xuebao* 國文學報 49 (2011), pp. 135–160.

Chiang, Bian: "Market Price, Labor Input, and Relation of Production in Sarawak's Edible Birds' Nest Trade", in Eric Tagliacozzo and Wen-chin Chang (eds.), *Chinese Circulations: Capital, Commodities, and Networks in Southeast Asia* (Durham and London: Duke University Press, 2011), pp. 407–439.

Couvreur, Séraphin (tr.): *Mémoires sur les bienséances et les cérémonies*, 2 vols. (Paris: Cathasia, etc., 1930).

Cutter, Robert J.: *Cao Zhi (192–232) and His Poetry* (University of Washington, 1983; unpubl. PhD dissertation).

Cutter, Robert J.: *The Brush and the Spur. Chinese Culture and Cockfight* (Hong Kong: The Chinese University Press, 1989).

Davis, A. R.: "Su Shih's 'Following the Rhymes of T'ao Yüan-ming's Poems': A Literary or a Psychological Phenomenon", *Journal of the Oriental Society of Australia* 10.1/2 (1975), pp. 93–108.

Davis, A. R.: *T'ao Yuan-ming, AD 365–427. His Works and Their Meaning* (Cambridge, etc.: Cambridge University Press, 2009).

Delacour, Jean, and Pierre Jabouille: *Les oiseaux de l'Indochine française*, 4 vols. (Paris: 1931).

Delacour, Jean, and Pierre Jabouille: "Oiseaux des îles Paracels", 3e mémoire of Armand Krempf (ed.), *Travaux du Service océanographique des pêches de l'Indochine* (Saigon: Fondation du Gouvernement Général de l'Indochine, 1930).

Dien, Dora Shu-fang: *Empress Wu Zetian in Fiction and in History: Female Defiance in Confucian China* (New York: Nova Science Publishers, 2003).

Du Halde, J. B.: *Description géographique, historique, chronologique, politique et physique de l'Empire de la Chine et de la Tartarie chinoise, enrichie des cartes générales et particulieres de ce Pays...*, 4 vols. (The Hague: Henri Scheurleer, 1736).

Egan, Romald C.: *Word, Image, and Deed in the Life of Su Shi* (Cambridge, Mass., etc.: Harvard University Press, 1994).

Ehmke, Eva: *Das Hai-cha yu-lu als eine Beschreibung der Insel Hainan in der Ming-Zeit* (Hamburg: Gesellschaft für Natur- und Völkerkunde Ostasiens, 1990).

Erkes, Eduard: "Der ikonographische Charakter einiger Chou-Bronzen. II: Die Eule – und Nachtrag...", *Artibus Asiae* 7.1–4 (1937), pp. 92–108, plus addendum in same 8.1 (1940), p. 49.

Étchécopar, R. D., et al.: *Les oiseaux de Chine, de Mongolie et de Corée non passereaux*, 2 vols. (vol. 1: Papeete, Tahiti: Les Éditions de Pacifique, 1978; vol. 2: Paris: Société Nouvelle des Éditions Boubée, 1983).

Fan Chengda (author), James M. Hargett (tr.): *Treatises of the Supervisor and Guardian of the Cinnamon Sea* (Seattle and London: University of Washington Press, 2010).

Fan Limei 范麗梅: "Yuedu 'chun zhi benben': *Shi jing* yinyong yu zhujie de duoceng quanshi" 閱讀《鶉之奔奔》：詩經引用與注解的多層詮釋, *Journal of Chinese Studies* 中國文化研究所學報 58 (January 2014), pp. 1–38.

Franke, Wolfgang (orig. author), and Liew-Herres Foon Ming (ed. of enlarged version): *Annotated Sources of Ming History. Including Southern Ming and Works on Neighbouring Lands, 1368–1661*, 2 vols. (Kuala Lumpur: University of Malaya Press, 2011).

Fry, C. Hilary, Kathie Fry (text), Alan Harris (illustrations): *Kingfishers, Bee-eaters and Rollers. A Handbook* (Princeton: Princeton University Press, 1992).

Goodrich, L. Carrington, and Chaoying Fang (eds.): *Dictionary of Ming Biography*, 2 vols. (New York and London: Columbia University Press, 1976).

Graham, A. C. (tr.): *The Book of Lieh-tzu* (London: John Murray, 1960).

Graham, William T.: "Mi Heng's 'Rhapsody on a Parrot'", *Harvard Journal of Asiatic Studies* 39 (1979), pp. 39–54.

Guangdong sheng kunchong yanjiusuo dongwushi, Zhongshan daxue shengwu xi 廣東省昆蟲研究所動物室，中山大學生物係: *Hainan dao de niao shou* 海南島的鳥獸 (Beijing: Kexue chubanshe, 1983).

Gudger, E. W.: "Fishing with the Cormorant. I. In China", *The American Naturalist* 60.666 (1926), pp. 5–41.

Guignard, Marie-Claude: *Aufzeichnungen über die Wunder des Südens. Übersetzung und Interpretation des Lingbiao-luyi von Liu Xun* (Universität Hamburg 1982; unpubl. MA thesis).

Guisso, Richard W. L.: *Wu Tse-t'ien and the Politics of Legitimation in T'ang China* (Bellingham: Western Washington Univ. Program..., 1978).

Guo Fu 郭郛 et al.: *Zhongguo gudai dongwuxue shi* 中國古代動物學史 (Beijing: Kexue chubanshe, 1999).

Hainan dao de niao shou, see Guangdong sheng kunchong yanjiusuo dongwushi, Zhongshan daxue shengwu xi.

Han Xuehong 韓學宏: "Tang shi zhong de zhenqin shuxie" 唐詩中的珍禽書寫, *Changgeng renwen shehui xuebao* 長庚人文社會學報 1.7 (2014), pp. 21–48.

Han Xuehong (author), Yang Dongfeng 楊東峰 (illustrations), Yuan Xiaowei 袁孝維 (ed.): *Jing dian Tang shi niaolei tujian* 經典唐詩鳥類圖鑒 (Zhengzhou: Zhongzhou guji chubanshe, 2005).

Hargett, James M.: "Clearing the Apertures and Getting in Tune: The Hainan Exile of Su Shi (1037–1101)", *Journal of Sung-Yuan Studies* 30 (2000), pp. 141–167.

Hargett, James M. (tr.): *Treatises*. See Fan Chengda.

Heeren-Diekhoff, Elfie: *Das* Hsi-Ching-Tsa-Chi. *Vermischte Aufzeichnungen über die westliche Hauptstadt* (Munich: Druckerei Fischer, 1981).

Herrmann, Konrad (tr., ed.): *Shen Kuo: Pinsel-Unterhaltungen am Traumbach. Das gesamte Wissen des alten China* (Munich: Eugen Diederichs Verlag, 1997).

Hightower, James R.: "Chia I's Owl Fu", *Asia Major* 7 (1959), pp. 125–130.

Hinton, David (tr.): *The Selected Poems of T'ao Ch'ien* (Port Townsend, WA: Copper Canyon Press, 1993).

Hirth, Friedrich, and W. W. Rockhill (eds., trs.): *Chau Ju-kua: His Work on the Chinese and Arab Trade in the Twelfth and Thirteenth Centuries, Entitled Chu-fan-chï* (rpt. Taibei: Ch'eng-Wen Publishing Company, 1970).

Hoffmann, Alfred: "Vogel und Mensch in China", *Nachrichten der Gesellschaft für Natur- und Völkerkunde Ostasiens* 88 (1960), pp. 45–77.

Hu Fan 胡凡: *Jiajing zhuan* 嘉靖傳 (Beijing: Renmin chubanshe, 2004).

Idema, Wilt L.: "The Filial Parrot in Qing Dynasty Dress: A Short Discussion of the Yingge baoju an [Precious Scroll of the Parrot]", *Journal of Chinese Religions* 30 (2002), pp. 77–96.

Jordan, J. N.: "Su Tung-p'o in Hainan", *The China Review* 12.1 (1883), pp. 31–41.

Kaiser, Thomas: "Unsterblich problematisch: *Grus japonensis*", in Ptak (ed.), *Tiere im alten China*, pp. 3–16.

Knechtges, David R. (tr.): *Wen xuan or Selections of Refined Literature*, 3 vols. (Princeton: Princeton University Press, 1982–1996).

Kroll, Paul W.: "Seven Rhapsodies of Ts'ao Chih", *Journal of the American Oriental Society* 120.1 (2000), pp. 1–12.

Kroll, Paul W.: "The Egret in Medieval Chinese Literature", *Chinese Literature: Essays, Articles, Reviews* 1.2 (1979), pp. 181–196.

Kroll, Paul W.: "The Image of the Halcyon Kingfisher in Medieval Chinese Poetry", *Journal of the American Oriental Society* 104.2 (1984), pp. 237–251.

Kurz, Johannes L.: *Das Kompilationsprojekt Song Taizongs (reg. 976–996)* (Bern, etc.: Peter Lang, 2003).

Lai, C. M. Lai: "Avian Identification of *jiu* 鳩 in the *Shijing*", *Journal of the American Oriental Society* 117.2 (1997), pp. 350–352.

Lai, C. M. Lai: "Messenger of Spring and Morality: Cuckoo Lore in Chinese Sources", *Journal of the American Oriental Society* 118.4 (1998), pp. 530–542.

Laufer, Berthold: *The Domestication of the Cormorant in China and Japan* (Chicago: Field Museum of Natural History, 1931; several reprints).

Legge, James: *The Chinese Classics*. Vol. 4: *The She King* (rpt. Taibei: Wen shi zhe chubanshe, 1971).

Lei Fumin 雷富民 and Lu Taichun 盧汰春: *Zhongguo niaolei teyou zhong* 中國鳥類特有種 (Beijing: Kexue chubanshe, 2006).

Leijon, Peer-Olow: "'Shooting Orioles', A Painting Signed [by] Ma Yüan", *Bulletin of the Museum of Far Eastern Antiquities* 49 (1977), pp. 17–29.

Leimbiegler, Peter: *Mei Yao-ch'en (1002–1060). Versuch einer literarischen und politischen Deutung* (Wiesbaden: Harrassowitz Verlag, 1970).

Li Changzhi 李長之: *Tao Yuanming zhuan lun* 陶淵明傳論 (Tianjin: Tianjin renmin chubanshe, 2007).

Li Haixia 李海霞: "Daxing cidian niao shou citiao shiyi jiubu" 大型詞典鳥獸詞條釋義糾補 *Sanxia daxue xuebao* (*renwen shehui kexue ban*) 三峽大學學報 (人文社會科學版) 30.1 (2008), pp. 61–64.

Li Mo 李默: *Guangdong fangzhi yaolu* 廣東方志要錄 (Guangzhou: Guangdongsheng difangzhi bianzuan weiyuanhui bangongshi, c. 1987).

Li Xiangtao 李湘濤: *Gamebirds of China* (Beijing: Zhongguo linye chubanshe, 2004).

Li Xiangtao: *Raptors of China* (Beijing: Zhongguo linye chubanshe, 2004).

Li Xiaoming 李曉明 and Ma Yiqing 馬逸清: *Dandinghe yanjiu* 丹頂鶴研究 (Shanghai: Keji jiaoyu chubanshe, 2002).

Lin Tianwei (Lin Tien-wai) 林天蔚: *Songdai xiangyao maoyi shigao* 宋代香藥貿易史稿 (Hong Kong: Zhongguo xueshe, 1960).

Lin Yutang: *The Gay Genius: The Life and Times of Su Tungpo* (London and Toronto: William Heinemann, 1948).

Liu Mau-tsai: *Kutscha und seine Beziehungen zu China vom 2. Jh. v. Chr. bis zum 6. Jh. n. Chr.*, 2 vols. (Wiesbaden: Otto Harrassowitz Verlag, 1969).

Mackerras, Colin (ed., tr.): *The Uighur Empire according to the T'ang Dynastic Histories: a Study in Sino-Uighur Relations 744–840* (Canberra: Australian National University Press, 1972).

Manzi, Maya, and Oliver T. Comes: "Cormorant Fishing in Southwestern China: A Traditional Fishery under Siege", *The Geographical Review* 92.4 (2002), pp. 597–603.

McCraw, David R.: "Along the Wutong Trail: The Paulownia in Chinese Poetry", *Chinese Literature: Essays, Articles, Reviews* 10.1/2 (1988), pp. 81–107.

Meyr, Anna: *Der Kuckuck im alten China. Von den Anfängen bis zur Song-Zeit* (Ludwig-Maximilians-Universität, 2010; unpubl. MA thesis).

Mittag, Achim: "Becoming Acquainted with Nature from the *Odes*: Sidelights on the Study of the Flora and Fauna in the Song Dynasty's *Shijing* 詩經 (Classic of Odes) Scholarship", in Hans-Ulrich Vogel and Gunter Dux (eds.), *Concepts of Nature: A Chinese-European Cross-cultural Perspective* (Leiden, etc.: Brill, 2010), pp. 310–344.

Mote, Frederick W., and Denis Twitchett (eds.): *The Cambridge History of China. Vol. 7: The Ming Dynasty, 1368–1644.* Part I (Cambridge, etc.: Cambridge University Press, 1988).

Mühlbauer, Andrea A.: *Greifvögel und Beizjagd in Asien* (Ludwig-Maximilians-Universität, 2014; unpubl. MA thesis).

Nappi, Carla: *The Monkey and the Inkpot. Natural History and Its Transformation in Early Modern China* (Cambridge, Mass.: Harvard University Press, 2009).

Netolitzky, Almut: *Das Ling-wai tai-ta von Chou Ch'ü-fei. Eine Landeskunde Südchinas aus dem 12. Jahrhundert* (Wiesbaden: Franz Steiner Verlag, 1977).

Nienhauser, William H.: *An Interpretation of the Literary and Historical Aspects of the* Hsi-ching tsa-chi *(Miscellanies of the Western Capital)* (Indiana University, 1972; unpubl. PhD dissertation).

Ogilvie-Grant, William R.: "On the Birds of Hainan", *Proceedings of the Zoological Society of London* (1900), pp. 475–504.

Pelliot, Paul: *Mémoire sur les coutumes du Cambodge de Tcheou Ta-Kouan. Version nouvelle, suivie d'un commentaire inachevé* (Paris: Librairie d'Amérique et d'Orient Adrien Maisonneuve, 1951; originally in *Bulletin de l'École française d'Extrême-Orient* 2, 1902).

Pohl, Karl-Heinz Pohl (tr.): *Der Pfirsichblütenquell: Gesammelte Gedichte* (Cologne: Eugen Diederichs, 1985).

Ptak, Roderich: "Asian Trade in Cloves circa 1500: Quantities and Trade Routes – A Synopsis of Portuguese and Other Sources", in Francis A. Dutra and João Camilo dos Santos (eds.), *Proceedings of the International Colloquium on the Portuguese and the Pacific. University of California, Santa Barbara, October 1993*, Publications of the Center of Portuguese Studies 10 (Santa Barbara: Center for Portuguese Studies, University of California, Santa Barbara, 1995), pp. 149–169.

Ptak, Roderich: *Birds and Beasts in Chinese Texts and Trade* (Wiesbaden: Harrassowitz Verlag, 2013).

Ptak, Roderich: "China and the Trade in Cloves, circa 960–1435", *Journal of the American Oriental Society* 113.1 (1993), pp. 1–13.

Ptak, Roderich: "Chinese Bird Imports from Maritime Southeast Asia, c. 1000–1500", *Archipel* 84 (2012), pp. 197–245.

Ptak, Roderich: *Exotische Vögel: Chinesische Beschreibungen und Importe* (Wiesbaden: Harrassowitz Verlag, 2006), pp. 11–33.

Ptak, Roderich: "Notizen zum *Qinjiliao* 秦吉了 oder Beo (*Gracula religiosa*) in alten chinesischen Texten (Tang- bis mittlere Ming-Zeit)", *Monumenta Serica* 44 (2007), pp. 447–469.

Ptak, Roderich: "The Avifauna of Macau: A Note on the *Aomen jilüe*", *Monumenta Serica* 57 (2009), pp. 193–230.

Ptak, Roderich (ed.): *Tiere im alten China. Studien zur Kulturgeschichte* (Wiesbaden: Harrassowitz Verlag, 2009).

Ptak, Roderich: "Weiße Papageien' (*bai yingwu*) in frühen chinesischen Quellen", in same, *Tiere im alten China*, pp. 31–48.

Ptak, Roderich:: "Zhanghai: Raum und Konzept. Von den Anfängen bis zur Tang-Zeit", in Shing Müller et al. (eds.), *Guangdong: Archaeology and Early Texts. Archäologie und frühe Texte (Zhou-Tang)* (Wiesbaden: Harrassowitz Verlag, 2004), pp. 241–253.

Ptak, Roderich: "Zhuhai dongwu lishi de yi bi: Jiajing 'Xiangshan xianzhi' li de yulei yanjiu" 珠海動物歷史的一瞥：嘉靖《香山縣志》里的羽類研究, *Aomen ligong xuebao* 澳門理工學報 15 (2012), pp. 173–184.

Pu Youjun 蒲友俊: "Chaoyue kunjing: Su Shi zai Hainan" 超越困境：蘇軾在海南, *Sichuan shifan daxue xuebao (shehui kexue ban)* 四川師範大學學報 (社會科學版) 84.2 (1992), pp. 75–83.

Read, Bernard E.: *Chinese Materia Medica: Avian Drugs* (rpt. Taibei: Southern Materials Center, 1977; originally in *Peking Natural History Bulletin* 6.4, 1932).

Reed, Carrie E.: *A Tang Miscellany. An Introduction to* Youyang zazu (New York, etc.: Peter Lang, 2003).

Rhie, Marilyn M.: *Early Buddhist Art of China and Central Asia*. Vol. 1: *Later Han, Three Kingdoms and Western Chin in China and Bactria to Shan-shan in Central Asia* (Leiden, etc.: Brill, 2007).

Röder, Mathias: "Vom kopfüber Hängenden oder *daoguaniao*", in Ptak (ed.), *Tiere im alten China*, pp. 17–30.

Rothschild, Norman Harry: *Wu Zhao. China's Only Women Emperor* (New York: Pearson Longman, 2007).

Salmon, Claudine: "Le gout chinois pour les nids de salanganes et ses répercussions économiques en Indonesie (XVe/XVIe–XXIe s.)", *Archipel* 76 (2008), pp. 251–290.

Schafer, Edward H.: "Falconry in T'ang Times", *T'oung Pao* 46.3–5 (1958), pp. 293–338.

Schafer, Edward H.: "Notes on Duan Chengshi and His Writing", *Asiatische Studien* 16 (1963), pp. 14–34.

Schafer, Edward H.: "Parrots in Medieval China", in Søren Egerod and Else Glahn (eds.), *Studia serica Bernhard Karlgren dedicata. Sinological Studies Dedicated to Bernhard Karlgren on His Seventieth Birthday, October Fifth, 1959* (Copenhagen: Ejnar Munksgaard, 1959), pp. 271–282.

Schafer, Edward H.: *The Golden Peaches of Samarkand. A Study of T'ang Exotics* (Berkeley, etc.: University of California Press, 1963).

Schafer, Edward H.: *The Shore of Pearls* (Berkeley, etc.: University of California Press, 1969).

Schafer, Edward H.: *The Vermilion Bird. T'ang Images of the South* (Berkeley, etc.: California University Press, 1963).

Serruys, Henry: "A Note on the Names of the Hoopoe in Chinese and Mongol", *Canada-Mongolia Review* 3.2 (1977), pp. 110–117.

Shi Haitao 史海濤, Meng Jiliu 蒙激流 et al. (eds.): *Hainan luqi beizhui dongwu jiansuo* 海南陸栖背椎動物檢索 (Haikou: Hainan chubanshe, 2001).

Siebert, Martina: "Klassen und Hierarchien, Kontrastpaare und Toposgruppen: Formen struktureller Eroberung und literarischer Vereinnahmung der Tierwelt im alten China", *Zeitschrift der Deutschen Morgenländischen Gesellschaft* 162 (2012), pp. 171–196.

Siebert, Martina: *Pulu* 譜錄. *"Abhandlungen und Auflistungen" zu materieller Kultur und Naturkunde im traditionellen China* (Wiesbaden: Harrassowitz Verlag, 2006).

Simoons, Frederick J.: *Food in China: A Cultural and Historical Inquiry* (Boca Raton, etc.: CRC Press, 1991).

Situ Shangji 司徒尚紀 and Li Yan 李燕: "Zhengde 'Qiongtai zhi': yi bu jiechu de fangyu zhi zuo" 正德《瓊臺志》：一部傑出的方輿之作, in Zhou Weimin 周偉民 (ed.), *Qiong Yue difang wenxian. Guoji xueshu yantaohui lunwenji* 瓊粤地方文獻. 國際學術研討會論文集 (Haikou: Hainan chubanshe, 2002), pp. 91–97.

Song Dongliang 宋東亮 et al.: "Anchun de zhonglei, fenbu, tezheng ji jiazhi" 鵪鶉的種類、分布、特征及價值, *Anhui nongye kexue* 安徽農業科學 36.34 (2008), pp. 15010–15012.

Soymié, Michel: "Le Lou-feou chan: Étude de géographie religieuse", *Bulletin de l'École française d'Extrême-Orient* 48 (1954), pp. 1–139.

Spring, Madeline K.: *Animal Allegories in T'ang China* (New Haven: American Oriental Society, 1993).

Spring, Madeline K.: "The Celebrated Cranes of Po Chü-i", *Journal of the American Oriental Society* 111.1 (1991), pp. 8–18.

Stackmann, Ulrich: *Die Geschichte der chinesischen Bibliothek Tian Yi Ge vom 16. Jahrhundert bis in die Gegenwart* (Stuttgart: Franz Steiner Verlag, 1990).

Stercks, Roel: "Animal Classification in Ancient China", *East Asian Science, Technology and Medicine* 23 (2005), pp. 26–53.

Sterckx, Roel: *The Animal and the Daemon in Early China* (Albany: State University of New York Press, 2002).

Strassberg, Richard E. (ed., tr.): *A Chinese Bestiary: Strange Creatures from the Guideways Through Mountains and Seas* (Berkeley, etc.: University of California Press, 2002).

Stumpfeldt, Hans: *Ein Garten der Sprüche. Das Shuo-yüan des Liu Hsiang (79–8 v. Chr.)*, vol. 2 (Gossenberg: Ostasienverlag, 2011).

Sun Shu'an 孫書安 (ed.): *Zhongguo bowu bieming da cidian* 中國博物別名大辭典 (Beijing: Beijing chubanshe, 2000).

Sun Xinzhou 孫新周: "Chixiao chongbai yu Huaxia lishi wenming" 鴟鴞崇拜與華夏歷史文明, *Tianjin shifan daxue xuebao (shehui kexue ban)* 天津師範大學學報 (社會科學版) 5 (2004), pp. 31–37.

Swinhoe, Robert: "On the Ornithology of Hainan", *Ibis*, new ser. 6 (1870), pp. 77–97, 230–256, 342–367.

Tan Mei Ah [陳美亞]: "Beyond the Horizon of an Avian Fable: 'Dazui wu' as an Allegory of the Political Reforms of Wang Shuwen", *Journal of Chinese Studies* 中國文化研究所學報 51 (2010), pp. 217–252.

Taylor, Ian M.: "'Guan, guan' Cries the Osprey: An Outline of Pre-modern Chinese Ornithology", *Papers on Far Eastern History* 33 (1986), pp. 1–22.

Tu, Hsiao-Wei, and Lucia Liu Severinghaus: "Geographic Variation of the Highly Complex Hwamei (*Garrulax canorus*) Songs", *Zoological Studies* (Taibei) 43.3 (2004), pp. 629–640.

Unschuld, Paul U.: *Pen-ts'ao. 2000 Jahre traditionelle pharmazeutische Literatur Chinas* (Munich: Heinz Moos Verlag, 1973).

Upton, Beth A.: *A Study of the Avian Canon* (University of California, 1972; unpubl. MA thesis; not seen).

Wallace, Leslie V.: "Representations of Falconry in Eastern Han China (A.D. 25–220)", *Journal of Sport History* 39.1 (2012), pp. 99–109.

Wang Chao 王釗 and Shi Zhenqing 史振卿: "Shi shu Zhengde 'Qiongtai zhi' de xueshu jiazhi" 試述正德《瓊臺志》的學術價值, *Wenxue jie* 文學界 (11/2012), pp. 236–237.

Wang Huijun 王會均: *Hainan wenxian ziliao jianjie* 海南文獻資料簡介 (Taibei: Wen shi zhi chubanshe, 1983).

Wang Huijun: "Ming xiu 'Qiongzhou fuzhi' yanjiu" 明修《瓊州府志》研究, in Zhou Weimin 周偉民 (ed.), *Qiong Yue difang wenxian. Guoji xueshu yantaohui lunwenji* 瓊粵地方文獻. 國際學術研討會論文集 (Haikou: Hainan chubanshe, 2002), pp. 119–158.

Wang Qishan 王岐山: *Zhongguo de he, yangji he bao* 中國的鶴, 秧雞和鴇 (Nantou: Guoli fenghuang guniao yuan, 2002).

Wang Ting 王頲: "Feng sou li yu: haiwai zhenqin 'daoguaniao' kao" 鳳藪麗羽：海外珍禽《倒掛鳥》考, in his *Xiyu Nanhai shidi yanjiu* 西域南海史地研究, ser. Wen shi zhe yanjiu congshu (Shanghai: Shanghai guji chubanshe, 2005), pp. 111–128.

Wang Ting: "Shu shi qiu zhi – Tang, Song liang dai baixian xuyang ji qixi di kao" 蜀士求雉 – 唐、宋兩代白鷳畜養及棲息地考, in his *Gudai wenhuashi lunji* 古代文化史论集 (Shanghai: Shanghai guji chubanshe, 2007), pp. 74–91.

Waterbury, Florance: *Bird-deities in China*, suppl. vol. to *Artibus Asiae* 10 (1952).

Watters, Thomas: "Chinese Notions about Pigeons and Doves", *Journal of the North-China Branch of the Royal Asiatic Society*, new ser. 4 (December 1867), pp. 225–242.

Wen Huanran 文煥然: "Hainan sheng yixie difangzhi kao" 海南省一些地方志考, *Neimenggu daxue xuebao (shehui kexue ban)* 內蒙古大學學報 (社會科學版) (1/1992), pp. 123–125.

Wheatley, Paul: "Geographical Notes on Some Commodities Involved in Sung Maritime Trade", *Journal of the Malayan Branch of the Royal Asiatic Society* 32.3 (1959), pp. 1–140.

Wu Dachun 武大椿: "Douji shikao" 鬥鷄史考, in Zhang Zhongge and Zhu Xianhuang (eds.), *Zhongguo xumu shiliao ji*, pp. 293–294.

Wu Dachun 武大椿: *Zhongguo douji he zawen* 中國鬥鷄和雜文 (Huhehot: Yuanfang chubanshe, 2003).

Wu Jiayi 吳佳翼: "Nanfang xiao fenghuang – wuse que" 南方小鳳凰 – 五色雀, *Wen shi zhishi* 文史知識 (11/2003), pp. 99–103.

Wu Yifeng 吳儀鳳: *Yong wu yu xu shi – Han Tang qinniaofu yanjiu* 詠物與敘事 – 漢唐禽鳥賦研究 (Taibei: Huamulan wenhua chubanshe, 2007).

Xian Yuqing 冼玉清: "Su Shi yu Hainan dongwu" 蘇軾與海南動物, *Lingnan xuebao* 嶺南學報 9.1 (1948), pp. 105–124.

Xie Chengxia 謝成俠: "Zhongguo jizhong de lishi yanjiu" 中國雞種的歷史研究, in Zhang Zhongge and Zhu Xianhuang (eds.), *Zhongguo xumu shiliao ji*, pp. 284–292.

Xie Chengxia: "Zhongguo yang ge de lishi" 中國養鴿的歷史, in Zhang Zhongge and Zhu Xianhuang (eds.), *Zhongguo xumu shiliao ji*, pp. 297–305.

Yan Zhongwei 顏重威: *Shi jing li de niaolei* 詩經裡的鳥類 (Taizhong: Xiang yu wenhua, 2004).

Yang, Lihui, and Deming An, with Jessica Anderson Turner: *Handbook of Chinese Mythology* (Santa Barbara, California: ABC-CLIO, 2005).

Yao Pinwen 姚品文 and Long Xiangyang 龍祥洋: "Tang Xianzu yu Hainan dao" 湯顯祖與海南島, *Hainan shifan xueyuan xuebao* (*shehui kexue ban*) 海南師範學院學報 (社會科學版) 66 (4/2003), pp. 76–80.

Yu Changsen 喻常森: *Yuandai haiwai maoyi* 元代海外貿易 (Xi'an: Xibei daxue chubanshe, 1994).

Zhang Rong 張榮: "Yi fang Qingchao wenguan buzi de shenshi" 一方清朝文官補子的身世, *Xiandai kuaibao* 現代快報, 21.02.2011.

Zhang Shitai 張世泰, Feng Weixun 馮偉勛 and Ni Junming 倪俊明: *Guancang Guangdong difangzhi mulu* 館藏廣東地方志目錄 (Guangzhou: Guangdongsheng Zhongshan tushuguan lishi wenxianbu, 1986).

Zhang Shuoren 張朔人: *Mingdai Hainan wenhua yanjiu* 明代海南文化研究 (Beijing: Shehui kexue chubanshe, 2013).

Zhang Zhongge 張仲葛: "Luci xiaoshi" 鸕鷀小史, in Zhang Zhongge and Zhu Xianhuang (eds.), *Zhongguo xumu shiliao ji*, pp. 306–308.

Zhang Zhongge: "Woguo jiaqin (ji, ya, e) de qiyuan yu xunhua de lishi" 我國家禽（雞、鴨、鵝）的起源與馴化的歷史, in Zhang Zhongge and Zhu Xianhuang (eds.), *Zhongguo xumu shiliao ji*, pp. 266–283.

Zhang Zhongge and Zhu Xianhuang 朱先煌 (eds.): *Zhongguo xumu shiliao ji* 中國畜牧史料集 (Beijing: Kexue chubanshe, 1986).

Zhao Zhengjie 趙正階: *Zhongguo niaolei zhi* 中國鳥類志 (*A Handbook of the Birds of China*). 2 vols. (Changchun: Jilin kexue jishu chubanshe, 2001).

Zheng Guangmei 鄭光美 and Zhang Cizu 張詞祖: *Birds in China* (Beijing: Zhongguo linye chubanshe, 2002).

Zheng Zuoxin 鄭作新 (=Cheng Tso-hsin): *A Synopsis of the Avifauna of China* (中國鳥類區系綱要) (Beijing: Science Press; Hamburg and Berlin: Paul Parey Scientific Publishers, 1987).

Zheng Zuoxin (ed.), *Zhongguo dongwu tupu: niaolei* 中國動物圖譜：鳥類 (3rd. ed. Beijing: Kexue chubanshe, 1987).

Zheng Zuoxin: *Zhongguo jingji dongwu zhi: niao lei* 中國經濟動物志：鳥類 (*Economic Birds of China*) (Beijing: Kexue chubanshe, 1993; originally 1963).

Zheng Zuoxin: *Zhongguo niaolei xitong jiansuo* 中國鳥類系統檢索 (Beijing: Kexue chubanshe, 1966).

Zheng Zuoxin: *Zhongguo niaolei zhong he yazhong fenlei minglu daquan* 中國鳥類種和亞種分類名錄大全 (*A Complete Checklist of Species and Subspecies of the Chinese Birds*) (Beijing: Kexue chubanshe, 1994).

Zhongguo yesheng dongwu baohu xiehui 中國野生動物保護協會, Qian Yanwen 錢燕文 (ed.): *Zhongguo niaolei tujian* 中國鳥類圖鑒 (*Atlas of Birds of China*) (Zhengzhou: Henan kexue chubanshe, 1995).

Zhou Yunzhong 周運中: "Hetuo yu Alabo ma" 鶴鴕與阿拉伯馬, in Shi Ping 時平 et al. (eds.), *Zhongguo hanghai wenhua luntan* 中國航海文化論壇, vol. 1 (Beijing: Haiyang chubanshe, 2011), pp. 307–313.

Zhu Xi 朱曦 and Zou Xiaoping 鄒小平: *Zhongguo lulei* 中國鷺類 (Beijing: Zhongguo linye chubanshe, 2001).

Zhu Xueqin 朱學勤: *Zhongguo huangdi huanghou baizhuan: Ming Shizong* 中國皇帝皇后百傳：明世宗 (Huhehot: Yuanfang chubanshe, 2002).

Zhu Yihui 朱逸輝: *Hainan mingren zhuanlue* 海南名人傳略, 2 vols. (Guangzhou: Zhongshan daxue chubanshe, 1992).

Index

Bird names in Chinese (small initials for individual kinds / species, large initials for families), English (small initials) and Latin; place names, ancient texts and some authors. Not listed: frequently-cited authors / titles (Li Shizhen, *Taiping yulan*, *Guangdong xinyu*, local chronicles). – Pinyin transcriptions of Chinese bird names with two characters usually written together (for example, *baihe* 白鶴); with more than one character written in segmented form (for example, *baifu yao* 白腹鷂), but there are rare exceptions.

Maritime Asia

Edited by Roderich Ptak, Thomas O. Höllmann, Jorge Flores and Zoltán Biedermann

---

Reihentitel bis Band 17: South China and Maritima Asia

21: Roderich Ptak (Hg./Ed.)

Marine Animals in Traditional China: Studies in Cultural History

Meerestiere im traditionellen China:
Studien zur Kulturgeschichte

2010. XXVIII, 154 pages, 16 ill., hc
ISBN 978-3-447-06421-7   € 44,– (D)

22: Roderich Ptak

Birds and Beasts in Chinese Texts and Trade

Lectures Related to South China
and the Overseas World

2011. XI, 140 pages, 10 plates, hc
ISBN 978-3-447-06449-1   € 38,– (D)

23: Shi Ping, Roderich Ptak (Hg.)

Studien zum Roman *Sanbao taijan Xiyang ji tongsu yanyi*

Band 1

2011. 196 Seiten, 5 Abb., gb
ISBN 978-3-447-06523-8   € 44,– (D)

24: Shi Ping, Roderich Ptak (Hg.)

Studien zum Roman *Sanbao taijian Xiyang ji tongsu yanyi*

Band 2

2013. 168 pages, 5 ill., hc
ISBN 978-3-447-10003-8   € 36,80 (D)

25: Zoltán Biedermann

The Portuguese in Sri Lanka and South India

Studies in the History of Diplomacy,
Empire and Trade, 1500–1650

2014. X, 205 pages, 19 ill., hc
ISBN 978-3-447-10062-5   € 56,– (D)

Zoltán Biedermann's book explores the Portuguese presence in Sri Lanka and South India with an emphasis on connections, interactions and adaptations. An introduction, six freshly revised case studies and an afterword provide historical insights into the making of Portuguese power in the region and point out new ways forward in the study of the subject. Themes explored include Portuguese diplomacy in Asia, the connected histories of Portugal, Sri Lanka and the Habsburg Empire, the importance of cartography for the development of Iberian ideas of conquest, the political mechanisms that allowed for the incorporation of Sri Lanka into the Catholic Monarchy of Philip II, and the remarkable resilience of elephant hunting and trading activities in Ceylon during the 17th and 18th centuries. A long chapter delves into the comparative urban histories of Portuguese and Dutch colonial ports in South Asia and reveals intriguing connections between colonialism, local identities and cosmopolitan attitudes. Taken together, the essays in this book question simplistic contrasts between Europe and Asia as well as between the Portuguese and the Dutch empires. *The Portuguese in Sri Lanka and South Asia* highlights the complex connections between the global and the local in early modern European-Asian interactions.

HARRASSOWITZ VERLAG · WIESBADEN
www.harrassowitz-verlag.de · verlag@harrassowitz.de

Maritime Asia

Edited by Roderich Ptak, Thomas O. Höllmann, Jorge Flores and Zoltán Biedermann

26: Jiehua Cai

Das *Tianfei niangma zhuan* des Wu Huanchu

2014. X, 212 Seiten, 14 Abb., 3 Tabellen, br
ISBN 978-3-447-10136-3   *€ 58,– (D)*

Das *Tianfei niangma zhuan* 天妃娘媽傳 des Wu Huanchu 吳還初 ist ein kurzer Roman aus der späten Ming-Zeit (1368–1644) über die Wundertaten der Göttin Tianfei, die – zumeist unter dem Namen Mazu 媽祖 – bis in die Gegenwart entlang den chinesischen Küsten und weit darüber hinaus als Schutzpatronin der Seefahrer verehrt wird. Die Erzählung berichtet von zwei bösen Dämonen – einem Affen- und einem Krokodilgeist –, die aus dem Himmel fliehen. Das veranlasst die mitfühlende Tochter eines Sternenfürsten, persönlich in die Welt der Menschen hinabzusteigen, um Unheil abzuwenden und für Ordnung zu sorgen. Sie wird dazu in die Familie Lin 林 geboren, steigt schließlich als Tianfei erneut in den Himmel auf und jagt die beiden Flüchtigen, auch jenseits der Reichsgrenzen, bis sie diese nach zahlreichen Abenteuern zur Strecke bringt.

In einer ausführlichen Studie der Hauptfiguren der Erzählung wird der langen Tradition dieser Götter und Monster in der chinesischen Literaturgeschichte nachgegangen, um so dem Schaffen des Verfassers näher auf die Spur zu kommen und die Feinheiten des Romans auskosten zu können. Auf der Grundlage dieser Detailstudien werden abschließend strukturelle und inhaltliche Deutungsansätze geboten, welche die religionshistorische Bedeutung des *Tianfei niangma zhuan* unterstreichen.

27: Claudine Salmon

Ming Loyalists in Southeast Asia

As Perceived through Various Asian and European Records

2014. Ca. XXXII, 134 pages, hc
ISBN 978-3-447-10272-8   *Ca. € 48,– (D)*

In order to attain a panoramic view of the Ming loyalists who, after the rise of the Qing in 1644, settled in various port cities of what is today Vietnam and the Malay World, this study seeks to make use of various European and Asian sources such as travelogs, commercial records, genealogies, inscriptions, old maps, drawings, and archaeological findings as well as researches focused on a few leading Chinese figures.

The 1st chapter, after analyzing how some European observers perceived the Ming refugees, tries to provide a diachronic view of the various migratory waves lasting for more than 40 years, and their geografical distribution. The 2nd chapter emphasizes some communities for which the documentation is rather rich to investigate the strategies used by the loyalists to adjust to the local environment. This includes the part their leaders played in the host countries, their contribution to urbanization, their attempts to solve local political issues, their role in the development of agriculture and Asian maritime trade, thanks to their close contacts with the Zheng state in Taiwan, but also to their connections with European companies. The 3rd and last chapter considers the eventual posthumous fame of these loyalists within Southeast Asia and abroad, especially in Europe during the 18th century and even later, and more recently in the Malay Peninsula, Singapore, and in South China, where some loyalist heroes have been given pride of place in local history, and also in historical fiction.

HARRASSOWITZ VERLAG · WIESBADEN
www.harrassowitz-verlag.de · verlag@harrassowitz.de